IMAGES
of America

Fire Lookouts of Glacier National Park

Ranger Frank Liebig at Swiftcurrent Pass. This photograph, taken sometime between 1905 and 1910, shows forest ranger Frank Liebig at Swiftcurrent Pass in what later became Glacier National Park. Liebig, who watched for fires from passes and mountaintops, was not only Glacier's first ranger but also its first firewatcher. (Courtesy of the National Park Service.)

On the Cover: Reynolds Ridge Lookout, which was under construction in this photograph taken in 1931, was first occupied and used in the summer of 1932. More information about the Reynolds Ridge Lookout can be found in chapter four. (Courtesy of Glacier National Park.)

IMAGES
of America

FIRE LOOKOUTS OF GLACIER NATIONAL PARK

David R. Butler

ARCADIA
PUBLISHING

Copyright © 2014 by David R. Butler
ISBN 978-1-4671-3114-8

Published by Arcadia Publishing
Charleston, South Carolina

Printed in the United States of America

Library of Congress Control Number: 2013945671

For all general information, please contact Arcadia Publishing:
Telephone 843-853-2070
Fax 843-853-0044
E-mail sales@arcadiapublishing.com
For customer service and orders:
Toll-Free 1-888-313-2665

Visit us on the Internet at www.arcadiapublishing.com

To Janet and Will Butler, whose support and love—despite my fixation with Glacier National Park—sustain me.

Contents

ACKNOWLEDGMENTS

Partial funding for research related to *Fire Lookouts of Glacier National Park* was provided by a Faculty Research Enhancement Program Grant from Texas State University, a contract administered by Dr. Dan Fagre with the Biological Resources Division of the US Geological Survey, and a grant from the Charles Redd Center of Brigham Young University.

Dr. Wendy Bigler, Dr. Janet Butler, William Butler, Dr. Dave "Rocky" Cairns, Dr. Lisa DeChano, Dr. George Malanson, Dr. Carol Sawyer, James Dietrich, William Marienau, Leslie Kerr, and the late Rich Flowers accompanied me on my forays to the lookouts and lookout sites. Janet and Will Butler also provided moral support and love throughout.

Fritz Klasner, formerly of the Biological Resources Division of the U.S. Geological Survey's Glacier Field Station, West Glacier, was very helpful in tracking down old photographs and historical details concerning the fire lookouts. Brian Kennedy, former editor and publisher of the *Hungry Horse News* in Columbia Falls, Montana, was also helpful in tracking down and providing photographs previously published in that newspaper. Thanks also to the several individuals who provided photographs of lookouts from their personal collections, including Tim Anderson from the Glacier Mountaineering Society, Brian Bolster, Dr. Lisa DeChano, the Glacier County Honey Co., Dr. Jon Kedrowski, Kari Nielsen, Shannon Paprocki, Dr. Lynn Resler, Bob Sihler, Dr. Melanie Stine, and Dr. Clayton Whitesides. Special thanks also go to the staff of the Glacier National Park archives, including the wonderful museum curator Deirdre Shaw and archive workers Ann Fagre, Jean Tabbert, and Somer Hileman. Hileman was especially helpful in ensuring that I received high-quality photographic reprints with accompanying historical information. Finally, my brother Mike Butler was extremely helpful in providing advice about writing books for Arcadia Publishing, and my editor Laura Bruns was wonderful to work with.

The abbreviated annotations used in courtesy lines for this book's images are as follows: US Geological Survey (USGS), the US National Park Service (NPS), the US Forest Service (USFS), Glacier National Park (GNP), and the *Hungry Horse News* of Columbia Falls, Montana (HHN). All unattributed photographs are from the author's personal collection.

Introduction

The mental image of an out-of-control wildfire sweeping through a virgin landscape sends involuntary shivers down the spines of many people. Recent summers in Glacier National Park (GNP), located in Montana, reinforce this image, with widespread fires in the park in the summers of 2001, particularly the Moose Fire; 2003, when more than 10 percent of the park was aflame and West Glacier and Apgar were temporarily evacuated for extended periods; and 2006, when a great portion of the lower St. Mary and Red Eagle Valleys burned. In 2007, smoke from the Skyland Fire immediately south of the park near US Highway 2 at Marias Pass forced the temporary closing of park access along Highway 2.

Most residents of the United States born before 1970, conditioned by the extraordinary effectiveness of the US Forest Service's Smokey Bear campaign ("Only YOU can prevent forest fires!"), were brought up thinking that wildfires were evil and destructive agents wreaking havoc on innocent mountains or rangeland. This philosophy, which viewed the forests as a timber resource to be protected for future use rather than as a natural habitat that should be allowed to be subjected to natural processes (including fire), drove firefighting policy throughout the United States for most of the 20th century. Even today, in spite of a growing recognition that wildfires introduce positive benefits to a landscape, people cringe when the evening news shows a "national treasure" such as Glacier or Yellowstone National Park "going up in flames."

In response to the widespread desire to control wildfires in the United States, thousands of fire lookout towers were constructed throughout the country, primarily by the US Forest Service (USFS) and the National Park Service (NPS), during the first half of the 20th century. Fire watchers were positioned in these structures on 24-hour alert, for seven days a week, during the season of high fire danger. If lookout towers did not exist in an area or had not yet been built, fire watchers were simply issued a tent by the agency for which they worked, such as the Forest Service or National Park Service, and sent to mountaintops to watch for fires. The value of permanent structures for fire observation became apparent in the 1920s and 1930s, and dozens of fire lookouts were constructed in and around Glacier National Park during this time. Cooperation between Glacier Park, the Flathead National Forest (FNF), and the Blackfeet Indian Reservation's Indian Service reached a peak in the mid-1930s and 1940s, when over two dozen fire lookouts in and adjacent to Glacier Park provided round-the-clock protection from forest fires.

From 1933 to 1938, the NPS collaborated with the USFS in a program initiated to obtain panoramic, 360-degree photographs from the 200 existing and proposed fire lookouts throughout the entire federal park system. Forest Service foresters also took panoramic photographs at all 1,400 fire lookouts and 200 patrol points in Oregon and Washington between 1933 and 1938. The photographic work for the National Park Service was carried out by junior forester Lester M. Moe, who was on loan from the Forest Service.

In the Glacier National Park region, Moe took panoramic photographs at 42 separate locations in GNP, the FNF, and the Blackfeet Reservation. The panoramic photography locations include

15 fire lookouts in the adjacent FNF that provided somewhat distant views into GNP, 10 roadside sites that provided detailed views of GNP terrain from ground level, several emergency-only lookouts that provided detailed photography from topographically high ground-level points or crudely constructed elevated platforms, and 16 fire lookouts within GNP or the adjacent Blackfeet Reservation.

Moe took nearly all of the photographs during the summer of 1935, although he took a few during the summer of 1937. Several of these photographs are reproduced in this book. Three 120-degree photographs were taken from each photography site (except for some roadside sites, which, if outside the park, only showed views into Glacier) with panchromatic film, and similar panoramas were also taken from many of the fire lookouts using black-and-white infrared film. The film was held stationary in an Osborne photo-recording transit while the lens rotated 120 degrees from right to left for each frame, providing a fixed focal length across the film. The sheer magnitude of Moe's efforts can only be appreciated when one realizes that an Osborne photo-recording transit weighed approximately 75 pounds and required supporting tripods that brought the total weight upward of 100 pounds, all of which had to be carried up to the fire lookouts, platforms, or topographic high points.

As technology changed in the latter half of the 20th century, and as attitudes toward wildfire changed among agencies such as the USFS and the NPS, fire lookout towers began to fall into disuse and disrepair as they were replaced by aerial observers in helicopters and planes. In Glacier National Park, the onset of aerial observations in 1954 sounded the death knell for many of the park's fire lookouts, and many were subsequently razed or removed. Each lookout had its own unique history, and that history was in danger of being lost.

Today, nine fire lookouts remain in Glacier Park, four or five of which are regularly manned during summer fire season. The ruins of two lookouts remain on the Blackfeet Reservation. In the Flathead National Forest, three lookouts that were part of the greater GNP area observation network remain. In this book, I have attempted to tell the story of all of Glacier Park's fire lookouts—those still remaining as well as those that no longer exist. Lookouts in the FNF and Blackfeet Reservation are also described, because they played an integral part in protecting the forests on the margins of the park. Photographs from a variety of sources reveal the rich history of the lookouts that protected Glacier from fires in the period prior to the mid-1950s and the lookouts still being manned today.

Fire lookouts are described by many writers as magical places, and are well-known as inspirational sites for writers and poets such as Jack Kerouac, Norman Maclean, and Gary Snyder, as well as environmental writers and naturalists such as Edward Abbey and Doug Peacock. They also serve as nostalgic, historical reminders of a simpler time before the Internet, wireless communication, and the widespread use of advanced technology for spotting and monitoring fire boundaries. This book offers a look at the history and geography of the historic fire lookouts of Glacier National Park and surrounding areas in northwest Montana. Upon finishing this book, I hope that you will feel knowledgeable about the fire lookouts of Glacier Park and want to visit them and help protect those that remain.

One

The Early Years

Glacier National Park was created by an act of Congress that was signed into law by Pres. William H. Taft on May 11, 1910. Prior to that, the area had been established as the Flathead Forest Preserve in 1897 and was under the direction of the newly created US Forest Service from 1905 to 1910. USFS policy was to attempt to squelch any and all fires as soon as possible, and this policy carried over into the new national park in 1910.

Early forest rangers covered wide areas of thousands to hundreds of thousands of acres, not only watching for and squelching fires but also patrolling for poachers and trappers. At first, fire watching simply involved climbing to an accessible high point and observing the surrounding terrain. Frank Liebig, who lived on the shore of Lake McDonald on the west side of the mountains, was the first forest ranger for what became Glacier National Park.

The summer of 1910 brought widespread fires throughout the northwest, including the Glacier National Park area. New park superintendent William Logan directed his small force of rangers as best as possible, but much of the area around West Glacier and Apgar burned. Recognizing the importance of watching for fires, permanent lookout points with map tables and adjacent cabins began to be established by the NPS in Glacier, and the USFS in the Flathead National Forest, throughout the 1910s and 1920s. Phone lines were eventually extended to these points, and in some cases, pole towers were constructed to offer better views of the surrounding forest. Pairs of fire watchers were dispatched across Glacier National Park and the surrounding Flathead National Forest thanks to important cooperation between the administrations of the NPS and the USFS. However, widespread fires in 1926 and 1929 illustrated the need for higher, more permanent lookout structures that would provide for around-the-clock observation of area forests throughout the summer fire season.

First Chief of the US Forest Service. Gifford Pinchot served as chief of the Division of Forestry from 1898 through 1905 and as the first chief of the US Forest Service from 1905 to 1910. Under Pinchot, the USFS hired the first rangers for the Flathead Forest Preserve in northwestern Montana. Several of these Flathead rangers subsequently became some of the first Glacier National Park rangers. (USFS.)

Early Glacier National Park Area Rangers. This April 1910 image shows a gathering of USFS rangers in northwest Montana. Frank Liebig (standing, second from left) is recognized as the first fire lookout/watcher in the area that became Glacier National Park. Liebig routinely climbed Stanton Mountain, which was adjacent to his cabin on the shores of Lake McDonald, to search for smoke that indicated a forest fire. Joe Cosley (standing, third from left), a famous park ranger in Glacier from 1910 to 1914, was fired in 1914 for trapping and poaching furs. (USFS.)

Frank Liebig, Early Ranger and Fire Watcher. Liebig poses here with trophy furs at one of his cabins. Early rangers were expected to hunt "varmints," including wolves, mountain lions, and bears. Liebig was assigned to guard the entirety of the northern two-thirds of what is now Glacier National Park. His assignment, according to a letter from Flathead Forest Preserve supervisor F.N. Haines, was the "whole country . . . from Belton [now West Glacier] to Canada and across the Rockies to the prairies between Waterton Lake and the foot of Saint Mary Lake." (GNP.)

William R. Logan, First Superintendent. Glacier National Park's first superintendent, William R. Logan (left), is shown at his camp at the mouth of Fish Creek in the park. The other men are chief clerk Clement S. Ucker (center), and W.W. Phillips. Logan became superintendent in 1910, the year of the park's founding, and was immediately tasked with managing the chaos of the massive fires of the summer of 1910. (USGS.)

SWIFTCURRENT VALLEY, 1911. This view shows the upper reaches of the Swiftcurrent Valley on the east side of the park, looking toward Swiftcurrent Peak (center) and Mount Wilbur (right). Broad swaths of burned forest from the summer of 1910 are visible on the forest floor. (USGS.)

Burned Forest and Many Glacier. In this photograph from 1920, burned forest is visible near the tree line in the foreground, while the center of the image shows the Many Glacier Hotel complex on the shore of Swiftcurrent Lake. A temporary fire lookout platform, known as the Alton Ridge Lookout, was constructed in the early 1930s to the right of the hotel complex. The platform lookout was removed sometime after 1937, which was the last year it was included on a map in the Glacier National Park archives. (USGS.)

Early Forest Service Fire Watchers. After the disastrous fires of 1910, fire watchers were dispatched across northwest Montana. For the first several years after 1910, lookouts consisted of nothing more than high points and a pair of binoculars, as shown here. Fire watchers such as ranger Griffin and forest guard Cameron, shown here, were typically sent out in pairs, with one man to locate the fire and observe it while the other rushed off to attempt to extinguish it. (USFS; photograph by W.J. Lubken.)

Forest Service Fire Watcher and a Telephone. By the 1920s, rudimentary telephones were installed at the bases of many lookout viewpoints in Glacier National Park and the Flathead National Forest. Men like the Flathead watcher shown here had to identify fire locations from high vantage points, then quickly hike down to the base of the mountain to use the emergency telephone. (USFS.)

Flathead National Forest Fire Watcher, 1921. By the mid-1920s, crude map tables and shelters with telephones began to be placed at more convenient, higher vantage points in order to expedite the call-in process. (USFS; photograph by F. Merrill.)

Riverview Mountain Viewed from Scalplock Mountain. The rounded peak in the right-center distance is Riverview Mountain in the southern area of the park. A crude platform lookout and cabin were constructed there in 1923 to replace a tent lookout placed that same summer. A phone line was also installed when the tent lookout was first established. The lookout was soon replaced by one with a more sweeping view from Scalplock Mountain, from which this photograph was taken. The Riverview Mountain Lookout was abandoned in the 1930s, and nothing remains of the structure.

Original Indian Ridge Fire Lookout in Glacier National Park. By the mid-1920s, the National Park Service began constructing more permanent log lookout structures in Glacier National Park. The first lookout at Indian Ridge in northwest Glacier National Park was a tent with a phone line built in 1923. Pictured here are the Indian Ridge pole lookout and cabin, which were built in 1925 and 1924, respectively. The people are, from left to right, "Slim the Guide," Alexander "Scotty" Beaton, and W. Burgman of Chicago. (GNP.)

Cabin-and-Roof Lookout at Huckleberry Mountain, 1926. Glacier National Park built a log cabin on Huckleberry Mountain with a cupola-style map stand on the roof (reached by ladder) in 1926. The cabin, located in the western part of the park, replaced an original map table with shelter lookout that was destroyed by fire in August 1926. The man in the doorway may be ranger Joe Heimes, who built the original map table and shelter. (GNP.)

Two

The First Wave of Present-day Glacier Park Lookouts

Widespread fires in 1926 and the large Half Moon Fire of 1929 convinced park personnel of the necessity for a broader network of fire lookouts within and adjacent to the park. Lookouts in the Blackfeet Indian Reservation (see chapter five), Flathead National Forest (see chapter six), and Glacier National Park were quickly envisioned and subsequently placed between 1929 and 1935. Ten park lookouts were constructed by the National Park Service in this period, six of which—those at Apgar, Huckleberry Mountain, Loneman Mountain, Mount Brown, Numa Ridge, and Scalplock Mountain—still exist. Bear Mountain, Curly Bear, Elk Mountain, and Reynolds Ridge Lookouts, also built during this period, were subsequently razed by the park (see chapter four). Several additional "emergency lookouts" on platforms and an existing building were also put into use in this period; all are marked on mid-1930s Glacier Park maps of fire lookout facilities.

Three lookouts were planned for the summer of 1929: Apgar, Loneman Mountain, and Mount Brown. Apgar and Loneman were both under construction when they were destroyed by the Half Moon Fire. Mount Brown is, therefore, the oldest permanent lookout structure in the park. Apgar and Loneman were quickly rebuilt, and these three lookouts share the same architectural style. The lookouts at Scalplock Mountain (1931), Huckleberry Mountain (1933), and Numa Ridge (1934) each have a similar style slightly different from the original three, possessing gentler roofs that overhang and protect the second-story catwalks. Roofs on Mount Brown, Apgar, and Loneman Mountain are more steeply pitched and do not overhang and shelter the catwalks. Among the destroyed lookouts, Bear Mountain was architecturally similar to the original three, whereas those at Curly Bear, Elk Mountain, and Reynolds Ridge had the gentler, overhanging roof design. The Huckleberry Mountain and Numa Ridge Lookouts replaced older, log-structure lookouts. The Huckleberry Mountain Lookout replaced an adjacent log cabin lookout with roof cupola and map stand. The Numa Ridge Lookout was built at a higher elevation, with a more expansive view, than the log-pole lookout lower on Indian Ridge in the northwest part of the park.

The 1929 Half Moon Fire. The 1929 Half Moon Fire in Glacier National Park—pictured here on August 21 from the Coram Ranger Station, about six miles southwest of the park—burned over 100,000 acres, with 50,000 acres of prime forest burned near park headquarters and around the lower end of Lake McDonald. This and other fires in 1929 spurred the development of permanent fire lookout structures in the park and adjacent national forest lands. (USFS; photograph by K.D. Swan.)

Panorama from Belton Point and Fire Effects. A 1935 panoramic photograph from the Belton Point Lookout shows some of the widespread effects of the 1929 Half Moon Fire in the Flathead National Forest. The words on the car door read "National Park Service Forestry." In 1935 and 1937, Lester M. Moe, of the US Forest Service, took panoramic photographs from every fire lookout in Glacier National Park and the Flathead National Forest, as well as from numerous roadside fire observation vantage points. (GNP; photograph by L.M. Moe.)

Material Packed for Building a Fire Lookout. In response to the disastrous fires of 1929, permanent fire lookout construction accelerated in both Glacier National Park and the Flathead National Forest in the early 1930s. Construction materials had to be carefully packed on well-trained mules or horses for transport to high vantage points. (USFS; photograph by K.D. Swan.)

Packer and Supply Train Heading to Lookout. Once permanent fire lookouts were constructed on peaks in remote locations, they had to be supplied every 10 to 14 days by packers leading mule trains. Supplies included food, water, mail, and the occasional treat. Famous Glacier fire lookout Leonard Stutsman, of Loneman Lookout, always eagerly anticipated a delivery of ice cream, which was packed in dry ice to survive the seven-mile trip. (USFS; photograph by K.D. Swan.)

Packer and Supply Train. The packers and their mule trains were lifelines for the fire watchers manning the fire lookouts. The fire watcher would typically provide the packer with a hot lunch in gratitude for receiving the needed supplies. (USFS.)

Mount Brown Lookout, 1982. The Mount Brown Lookout is one of three park lookouts that were under construction in the summer of 1929. The disastrous Half Moon Fire claimed the other two (Apgar and Loneman), making Mount Brown the oldest lookout in the park. It was completed in the late fall of 1929, after the fire was finally extinguished, and has the same basic design (a two-story cabin with catwalk and a short, steep hip roof that does not extend over the catwalk) as the present-day Apgar and Loneman Lookouts. The lower floor was used for storage, while the fire watcher lived in the glass-enclosed second floor. (NPS.)

Final Steps to Mount Brown Lookout. The Mount Brown Lookout sits atop a false summit west of the true peak of Mount Brown. The number on the roof is one of two originally placed on the roof in 1962 by the Montana Aeronautical Commission as part of a statewide plan to assist pilots with determining their locations (numbers were also placed on Huckleberry and Swiftcurrent Lookouts). (USGS.)

Inspection of Mount Brown Lookout, 1973. The Mount Brown Lookout was routinely manned through 1971. After then–park superintendent William Briggle inspected the lookout for safety issues in August 1973, it was subsequently used during fire emergencies in 1973 and 1974. Here, Superintendent Briggle, in cap, stands next to firewatcher Don Echlin and an Osborne Firefinder. (HHN; photograph by Mel Ruder.)

Mount Brown Lookout, 1998. This photograph shows the glaringly apparent need for restoration of the Mount Brown Lookout. The Montana Aeronautical Commission numbers from the roof were lying in a heap around the base of the lookout when the author visited it in 1998. Restoration work on the lookout began in 1999 and required an estimated $27,000.

View of Lake McDonald from Mount Brown Lookout. The Mount Brown Lookout provides a spectacular panorama of Lake McDonald, Glacier Park's longest and deepest lake. When visiting Mount Brown Lookout on a clear day, the Apgar and Huckleberry Lookouts are visible as dots on the southern and northern ends of the Apgar Range at the foot of Lake McDonald, and Loneman Lookout can be seen to the south. If the day is very clear, far to the northeast Swiftcurrent Lookout is barely visible on the crest of Swiftcurrent Mountain on the Continental Divide.

Resting after the Long Walk to Mount Brown Lookout. The trail to the Mount Brown Lookout, which starts at the Lake McDonald Lodge parking lot, climbs over 4,300 feet in elevation over the course of 5.4 miles. It is described in various hiking guides as "leg-burning," "one of the steepest and most grueling trails in the park," and "twenty-nine excruciating switchbacks." When hikers finally reach the lookout, a rest on the catwalk, such as is being enjoyed by David Cairns and Lisa DeChano in 1998, is certainly in order!

Apgar Lookout, 1982. Lookouts have been in place on Apgar Mountain since 1927, when watchers first used a lookout tent. This was doubtless a response to the widespread fires of 1926, which Glacier Park superintendent Charles Kraebel called "the worst fires since the terrible season of 1910." An elevated platform tower and cabin built in the spring of 1929 replaced the tent lookout, but these new structures were destroyed by the Half Moon Fire within two weeks of completion. Construction on the current lookout started in May 1930 and was completed prior to the fire season of 1930 at a cost of $1,600. (NPS.)

Apgar Lookout, 1998. Apgar Lookout was routinely manned until 1970, followed by a short period of usage during a fire emergency in 1972. The lookout was rehabilitated in the late 1990s. Although it is not currently in use, the lookout serves as an important communications center for park radio facilities, and visitors to the lookout will encounter numerous aerials and antennae, as shown here. Lisa DeChano is pictured in both photographs on this page.

Stairway to Upper Level of Apgar Lookout. Apgar Lookout is a National Park Service Standard Type 1 design (two-story, fully enclosed frame cabin, 14-by-14 feet, with living quarters and catwalk above and a cellar below) similar to the designs of the Mount Brown and Loneman Lookouts. Much of the overgrowth surrounding the lookout was burned off during the Robert Fire of 2003, leading to the reopening of the view.

Panoramic View from Apgar Lookout, 1935. The effects of the 1929 Half Moon Fire are readily apparent in this 1935 view from Apgar Lookout. The phone line linking the lookout to park headquarters stretches across the poles in the foreground. Almost the entire length of Lake McDonald is visible, as are the peaks beyond Snyder Ridge on the south side of the lake. A large

portion of the Livingston Range is visible to the north of Lake McDonald. To the southeast and south, the village of West Glacier and its golf course dominate the valley floor, surrounded by the Belton Hills, Desert Mountain, and the US Highway 2 region. (GNP; photograph by L.M. Moe.)

LONEMAN LOOKOUT AS VIEWED FROM HELICOPTER, 1995. Loneman Lookout sits atop Loneman Mountain, commanding a view of much of the southwestern quadrant of Glacier Park and the surrounding Flathead National Forest. This lookout is similar in design to Apgar and Mount Brown Lookouts.

LONEMAN LOOKOUT, 1949. Construction of a lookout on Loneman Mountain began in August 1929. While it was under construction, with the lookout about "one third up," the Half Moon Fire burned the completed framework and all the lumber that had been packed up to the site. Thanks to an emergency funding appropriation, work on the replacement lookout started in May 1930 and was finished in less than two weeks. (GNP; photograph by J.R. Hutchinson.)

View of Loneman Lookout with Shutters Down, 1982. With the window shutters down, Loneman Lookout appears to be a stark structure. The lookout was regularly manned through 1978 and for a short time during high fire danger in 1979. After it was rehabilitated in 2003, it had to be wrapped in fire-resistant material because of the Middle Fork Complex Fire. The lookout survived; in 2009, it was manned for the first time in 30 years. (NPS.)

Tourist Map from the 1960s Shows Loneman Lookout. In this 1960s tourist-oriented map that was sold throughout Glacier National Park, Loneman Lookout is memorialized as a landmark near US Highway 2 in southwest GNP. (Don Bloodgood.)

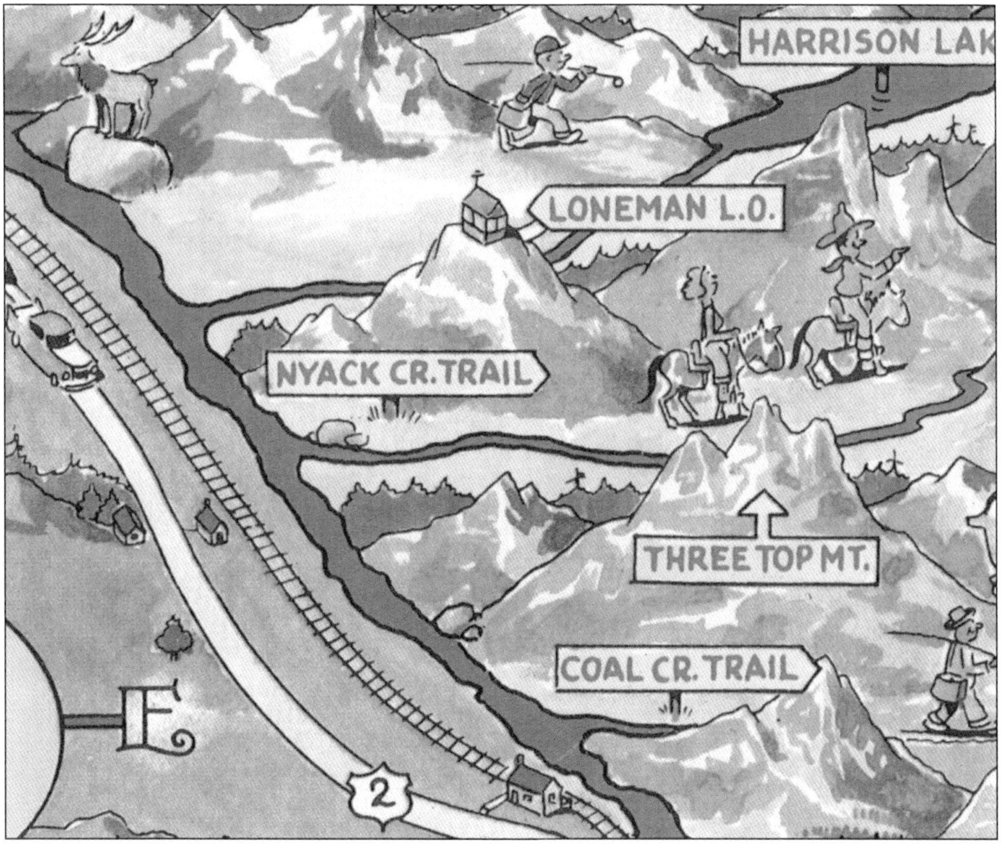

Loneman Lookout Interior. In this undated view of the shuttered interior, the fire watcher's bed is visible in the lower part of the image. On the table, behind the lantern, is the single most important feature in any fire lookout—the Osborne Firefinder. This device allowed a fire watcher to determine a fire's location with great accuracy by providing a compass bearing for the fire as well as an elevation relative to the fire lookout. (NPS.)

Undated View of Interior of Loneman Lookout Storage Room. The lower stories of two-story lookouts like Loneman were unfinished and used for storage of equipment as well as food. The windowless interiors helped keep food cool in an environment where electricity did not exist at the time of construction. Many park lookouts now have solar panels that provide electricity for refrigeration. (NPS.)

View of Great Northern Mountain from Loneman Lookout. Great Northern Mountain, located in the Great Bear Wilderness of the Flathead National Forest, dominates the view to the southwest. The US Geological Survey is conducting repeat photography from park fire lookouts to illustrate changes in the landscape since the time when Lester M. Moe took his panoramic photographs in 1935 and 1937. (USGS; photograph by Greg Pederson.)

CCC Men Arrive at Missoula Train Station, 1933. Civilian Conservation Corps workers arrived by train in Missoula, Montana, and were subsequently distributed throughout northwest Montana to provide needed labor for the construction of many trails and structures in the Flathead National Forest and Glacier National Park. Workers came from across the United States, with many hailing from large cities in the eastern United States. For such workers, the transition to the mountains of northwest Montana may have been startling. (USFS; photograph by K.D. Swan.)

CCC Men Heading to Job Site in Northwest Montana. Civilian Conservation Corps workers improved many trails in the Flathead National Forest and Glacier Park. They were also involved in the construction of some lookouts, including Curly Bear and Swiftcurrent. (USFS; photograph by K.D. Swan.)

CCC Camp at Many Glacier, Glacier National Park. These workers are pictured during a lunch break at the Civilian Conservation Corps Many Glacier camp in northeastern Glacier Park. (USFS; photograph by K.D. Swan.)

Burned Area at the Base of Scalplock Mountain, 1935. This photograph shows an area that burned in 1931 and that had also previously burned in 1910, near the base of Scalplock Mountain in southwestern Glacier National Park. A fire lookout was built above this site in the summer of 1931, a reaction to the regional 1929 fires and this 1931 fire. (GNP; photograph by L.M. Moe.)

Burned Areas at Base of Trail to Scalplock Lookout, 1935. This panoramic view looks west-southwest across the burned area at the base of Scalplock Mountain. The 1931 fire in this area helped spur the construction of Scalplock Lookout. Funding for its construction was transferred from money that had been marked for the building of a new Huckleberry Lookout in 1931; the need for Scalplock Lookout forced park officials to delay the Huckleberry Lookout's construction until 1933. (GNP; photograph by L.M. Moe.)

CCC Workers on the Trail to Scalplock Lookout, 1930s. Civilian Conservation Corps workers created the trail to Scalplock Lookout that still exists today. The trail climbs 2,900 vertical feet, over a series of 17 switchbacks, to the crest of the peak and the lookout at 6,919 feet. (USFS.)

Scalplock Lookout and Northward View to Mount St. Nicholas, 1975. Scalplock Lookout was completed in 1931 and is still manned, although it had a period of disuse during the 1970s. The lookout was not in use in 1975, when the author took this photograph. It was in somewhat rough condition (note the tilt of the stairs). Mount St. Nicholas dominates the skyline in this image.

Scalplock Lookout, 1982. By 1982, solar panels had been installed at Scalplock Lookout. Leonard Stutsman, the eponymous "lone man" of Loneman Lookout, manned Scalplock Lookout during his final years in Glacier National Park. Reclusive author and environmental activist Doug Peacock also spent several summers at Scalplock, as well as at Huckleberry Lookout. Note the improvement of the stairs since the 1975 photograph on the previous page. (NPS.)

Interior of Scalplock Lookout, 1993. Scalplock Lookout's interior is similar to those of other lookouts still in use in the park. The interior is spartan, as the primary focus of any lookout is the Osborne Firefinder in the center of the room. Note the glass insulators on the legs of the chair; this protection was necessary during lightning storms that frequent this and other mountaintops. (NPS; photograph by Jason Wilmot.)

Scalplock Lookout, 1998. In 1998, Scalplock Lookout was showing its age, but it was rehabilitated the following year. Modern solar panels now adorn the exterior of the lookout, and the overall exterior looks significantly better than it did in this photograph. Snow damaged the roof over the winter of 2011, but the damage was repaired that summer. The author is pictured here at the lookout.

Scalplock Lookout and Hitching Rack. The hitching rack illustrates the continuing importance of pack trains in supplying the still-active lookouts in Glacier Park, including Scalplock. In contrast to the roofs at Mount Brown, Apgar, and Loneman Lookouts, Scalplock Lookout has a gently sloping roof that extends over the second-floor catwalk. (Glacier County Honey Co.)

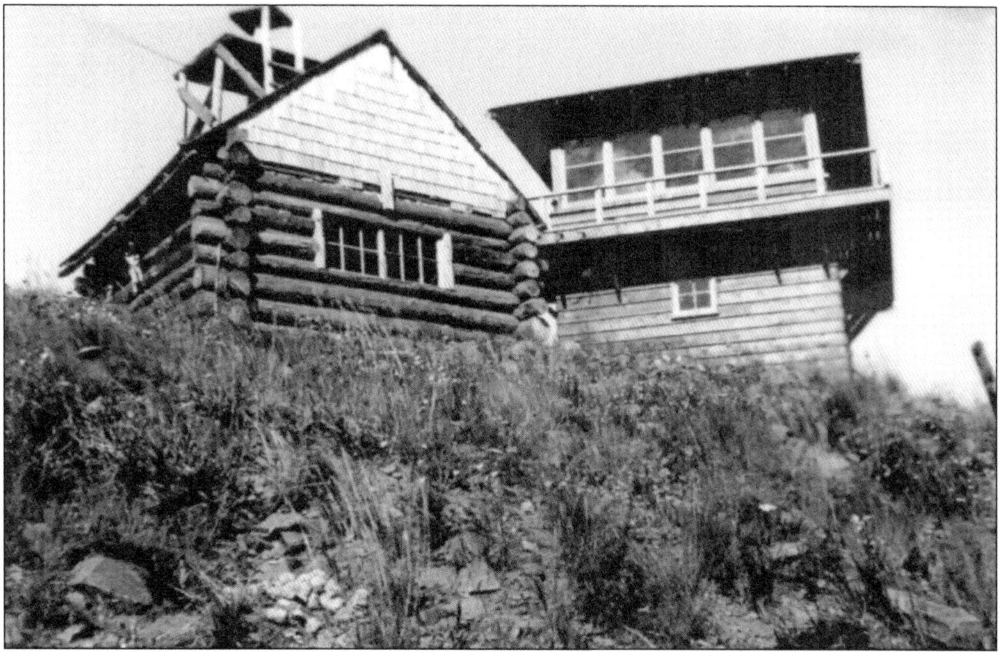

Old and New Huckleberry Lookouts, 1933. A new, standard two-story lookout was constructed on Huckleberry Mountain in 1933 at a cost of $1,600. The new lookout was adjacent to the old cabin-and-platform lookout constructed in 1926. The older lookout was erroneously said to have been destroyed by fire in 1929, but its continued use until 1933 allowed transference of the funds necessary to build Scalplock Lookout in 1931. (GNP.)

Dismantling the Old Huckleberry Lookout, 1940. By 1940, the old rooftop lookout was being dismantled. The man on the catwalk of the new lookout is unidentified. The lookout constructed in 1933 is a two-story, 14-by-14-foot structure with an extended roof over the catwalk similar to the one at Scalplock Lookout. This lookout still exists today. (GNP.)

Huckleberry Lookout, 1982. Huckleberry Lookout has been continuously manned since its construction in 1933. It survived a nearby fire in 1967 that burned over 8,000 acres in Glacier Park and the adjacent Flathead National Forest. Grizzly bear advocate and author Doug Peacock, who served as the model for the character George Hayduke in Edward Abbey's famous novel *The Monkey Wrench Gang*, was a fire watcher at Huckleberry in the early 1980s. (NPS.)

Interior of Huckleberry Lookout, 1982. This inventory photograph of Huckleberry Lookout reveals several updated appliances (compared to the "old days" of the wood stove). Bottled propane, which fuels the gas stove, must be brought in by packers. Out of frame, an Osborne Firefinder sits in the center of the lookout. (NPS.)

Packer Heading to Huckleberry Lookout, 1992. Even in modern times, pack trains are still a vital supply link for lookouts. Packers must deliver not only food, but also all water supplies for cooking and cleaning. (HHN; photograph by Tad Brooks.)

Fire Watcher and Child at Huckleberry Lookout, 1992 and 1998. In the image at right, fire watcher Chris Baker and her daughter Katie relax at the Huckleberry Lookout during the summer of 1992. Baker served as a fire watcher in Glacier Park from 1991 to 2004, spending most of her summers in the 1990s at Huckleberry. She served at the Swiftcurrent Lookout in the 2000s, when she was the fire watcher during the 2003 Trapper Fire. Baker worked as a firewatcher and ranger for a total of 16 summer seasons in the park. (Right, HHN; photograph by Tad Brooks; below, HHN; photograph by Emmett Berg.)

HUCKLEBERRY LOOKOUT FROM BELOW, 2012. This photograph, taken atop Huckleberry Mountain, looks much the same as photographs from any year between 1933—when the lookout was built—and today. The hitching rack and the "outhouse" are just over the crest of the mountain on the northeast side. (Kari Nielsen.)

Huckleberry Lookout Wrapped in Fire-retardant Material, 2001. In the late summer of 2001, the Moose Fire began in the Flathead National Forest. It eventually spread to over 60,000 acres, approximately one-quarter of which were within Glacier Park. The fire threatened the Huckleberry Lookout in a fashion similar to the fire of 1967. The park service wrapped the lookout in a protective insulating coat of reflective material, and the lookout survived. (NPS.)

Repeat Photography Project from Huckleberry Lookout. The US Geological Survey's repeat photography program, which is repeating the photographs taken by Lester M. Moe in 1935 and 1937, is designed to illustrate landscape changes in Glacier Park. In this July 2009 picture, USGS photographer Lindsey Bengston uses an original Osborne photo-recording transit, the same kind of camera used by Moe in the 1930s. (USGS.)

Huckleberry Lookout Interior and Catwalk. A view of the Osborne photo-recording transit on the catwalk of Huckleberry Lookout also offers a look at some of the kitchen utensils in the interior of the lookout. The repeat photographs taken in July 2009 will allow assessment of 74 years of landscape change since Lester M. Moe's original photographs, which he took in 1935 and 1937. (USGS; photograph by Lindsey Bengston.)

Trail to Huckleberry Lookout. This image offers a final view back to Huckleberry Lookout before heading down to the trailhead on the Camas Creek Road in Glacier Park. The hike is 12 miles round-trip and gains approximately 3,400 feet of elevation between the trailhead and the lookout. (Jon J. Kedrowski.)

Numa Ridge Lookout, 1934. Built in 1934 to replace the log structure called Indian Ridge Lookout, Numa Ridge Lookout has been in continuous service since its construction. It is a typical Glacier Park two-story, 14-by-14-foot frame cabin with an extended hip roof and a catwalk similar to those at Scalplock and Huckleberry Lookouts. (GNP.)

Insulator for Numa Ridge Phone Line. The Numa Ridge Lookout phone line is a historic structure listed in the National Register of Historic Places. This close-up shows a portion of the phone line with a ceramic insulator in a tree near the Bowman Lake Ranger Station. (NPS.)

SCOTTY BEATON (RIGHT) AND ANTON SCHOENBERGER. Beaton and his father-in-law, Schoenberger, lived in a North Fork cabin near Bowman Lake with Beaton's wife, Minnie. Beaton began working as the Indian Ridge fire watcher around 1916 and moved to Numa Ridge after its construction in 1934. Beaton retired in 1948, helped train a replacement in 1949, and died of pneumonia in 1950. (Sharon Lytle.)

NUMA RIDGE LOOKOUT IN 1982. Numa Ridge Lookout sits atop a low peak on the north shore of Bowman Lake at an elevation of 6,960 feet. The hike from Bowman Lake is about 11.5 miles round-trip, mostly through forest. (NPS.)

Numa Ridge Lookout, 1982. Numa Ridge Lookout served to help watchers spot fires and relay radio traffic during North Fork–area forest fires in 1998, 1999, and 2000. It also has expansive views of the Flathead National Forest. (NPS.)

View from Numa Ridge Lookout. Bowman Lake (at left) dominates the view from Numa Ridge Lookout. The small pond, known as Moose Pond, near center is passed (but typically not seen) on the hike up to the lookout from the foot of Bowman Lake. The outhouse facility for the lookout is at lower right. (Lynn M. Resler.)

PANORAMIC VIEW FROM NUMA RIDGE LOOKOUT, 1935. The head of Bowman Lake is not visible from Numa Ridge Lookout, but Rainbow Peak dominates the southern view. Peaks of the Whitefish

Range in the Flathead National Forest are visible in the distance at right. (GNP; photograph by L.M. Moe.)

Interior of Numa Ridge Lookout. The Osborne Firefinder dominates the center of the interior of Numa Ridge Lookout, as is appropriate. A propane stove is visible behind the Firefinder, as is a wood-burning stove that supplies heat. (Lynn M. Resler.)

"Abbey Lives" Graffiti on Numa Ridge Lookout Trail Sign. Environmentalist Edward Abbey, author of *The Monkey Wrench Gang* and numerous other books, and his wife were fire watchers at Numa Ridge in the summer of 1975. Abbey, a self-proclaimed "desert rat," wrote journal entries indicating that he did not particularly enjoy the cool, moist Glacier Park summer. This graffiti defacing the trail sign to Numa Ridge was photographed in the late 1990s.

Numa Ridge Lookout from Below, 1969. In 1969, Numa Ridge Lookout had not yet been modernized with solar panels. Beginning in the 1980s, solar panels began to become commonplace at lookouts. (NPS.)

Solar Panels on Numa Ridge Lookout, 2009. This picture of Numa Ridge Lookout was taken just slightly lower downhill than the previous photograph, with the same rocks visible in both. This modern view shows the solar panels that now adorn the flank of the lookout. (Jon J. Kedrowski.)

Three

On Top of the World

After 1935, the NPS began to construct fire lookouts on higher ground, either on top of a mountain peak (Swiftcurrent Lookout) or on high, rocky ridges (Porcupine Ridge and Heaven's Peak Lookouts)—in any case, they were placed well above the tree line in isolated locations. Vistas from these lookouts are largely unencumbered by peaks or surrounding vegetation. Swiftcurrent Lookout, constructed in 1936, sits at the highest elevation of all lookouts in the park. Porcupine Ridge Lookout, completed in 1939 on a long, rocky ridge high above Waterton Lake in the northern fringes of the park, was the last lookout completed before World War II. Porcupine Ridge Lookout has been called the most isolated lookout in Glacier Park and was last manned in the early 1970s.

During World War II, another isolated lookout was built when Mennonite conscientious objectors in the Civilian Public Service, an alternative to military service during the war, built Heaven's Peak Lookout for the NPS and Glacier National Park. This stone lookout was completed in September 1945 and only manned from 1946 to 1953. Swiftcurrent Lookout is still manned and maintained today, but both Heaven's Peak and Porcupine Ridge Lookouts underwent long periods of neglect and disuse before recent renovations.

Porcupine Ridge Lookout, near the Canadian border, serves as a communications relay outpost in the modern era of border security. The trail to Porcupine Ridge, also long disused, has been renovated to allow access to the communications devices at Porcupine Ridge.

No such luck for hikers who wish to visit Heaven's Peak Lookout. The trail to Heaven's Peak was begun in August 1939 under the auspices of the Works Progress Administration, and most of the trail was completed by November 1940, at which time WPA projects and camps withdrew from Glacier National Park. The final segment of the trail was built from June 1941 to August 1942 using Glacier Park allotments. After the lookout was abandoned in the 1950s, the trail also fell into disuse, and now hikers can only access Heaven's Peak Lookout via a long, off-trail day of bushwhacking.

Swiftcurrent Lookout on Swiftcurrent Mountain. Glacier Park's highest lookout, Swiftcurrent Lookout, is visible in this image as a small dot on the very top of Swiftcurrent Mountain (the rounded peak at left). The lookout commands an outstanding 360-degree view from its position well above the tree line. The lookout was completed in August 1936, but was not yet in use at the time of the spectacular Heaven's Peak fire at the end of that month. (USGS.)

Swiftcurrent Lookout from the Air, 1967. In this bird's-eye view, Swiftcurrent Lookout's position on the stony crest of its namesake mountain is visible at the end of the grueling trail leading up to it. As at Mount Brown and Huckleberry Mountain Lookouts, large numbers were placed on the roof in the 1960s as a navigational aid for low-flying aircraft. (NPS.)

Swiftcurrent Lookout from Ground Level, 1949. Swiftcurrent Lookout is an atypical, Swiss chalet–style structure atop a native stone base. A cellar door is visible at right. Because of its lofty position, a catwalk was not included in the lookout's construction, nor would a catwalk probably survive the strong winds atop the peak. Cables help hold the roof in place against the strong winds. (GNP; photograph by Henderson.)

North Face of Swiftcurrent Lookout, 1982. Shutters are numbered around the lookout, starting with the window adjacent to the front door. Each numbered shutter must be matched with its specific window for a perfect fit. (NPS.)

Front View of Swiftcurrent Lookout with Snow, 1982. The door of Swiftcurrent Lookout faces east, so the sun greets the fire watcher each morning unless the lookout is enshrouded in the clouds—a common occurrence at its elevation of 8,436 feet. In this photograph, taken in early summer, snow partially covers the front steps. (NPS.)

NORTHWARD VIEW FROM SWIFTCURRENT LOOKOUT, 2011. In this recent photograph, the northeast corner of the lookout is shown next to the spectacular view to the north. To the east is the Swiftcurrent Valley, and to the south are views of Mount Gould and the Logan Pass area. Heaven's Peak Lookout, not visible here, can be seen almost directly to the west from Swiftcurrent Lookout. (Melanie B. Stine.)

EASTWARD VIEW INTO SWIFTCURRENT VALLEY FROM SWIFTCURRENT LOOKOUT, 2011. A trip to the outhouse can be an exhilarating experience at Swiftcurrent Lookout! It is a good idea not to go beyond the facility, where a sheer cliff drop-off awaits. The long lake in the right background in this view is Lake Sherburne in the Swiftcurrent Valley. (Melanie B. Stine.)

Interior Damage from a Snowstorm at Swiftcurrent Lookout, 1993. Swiftcurrent Lookout was in almost-continuous summer use from its 1936 construction until it was damaged by a wind storm in the winter of 1991–1992. Shutters and windows were blown in, which allowed snow to accumulate; the snow subsequently melted and damaged the flooring, as shown here. The lookout was refurbished in 1999 and has been in continuous summer use since 2001. (NPS; photograph by Wayne Thompson.)

Shuttered Front of Swiftcurrent Lookout, 1998. In 1998, the lookout was still one year away from being refurbished. The stonework at the base of the lookout has held up extremely well since its construction. (Lisa M. DeChano.)

Rehabbed Front of Swiftcurrent Lookout, 2006. The Osborne Firefinder at the center of the lookout is visible through the front door of the rehabilitated structure. The lookout survived the 2003 Trapper Fire that burned through the tree line forest below; fire watcher Chris Baker provided radio updates when the view was not totally cloaked in smoke. (NPS.)

Welcoming Front of Swiftcurrent Lookout, 2011. In this summertime view Melanie Stine visits the lookout, with the Livingston Range beckoning to the west. Note that cables are still used to assist in keeping the roof in place against the peak's high winds. (Clayton J. Whitesides.)

Solar Panels at Swiftcurrent Lookout, 2011. Solar panels have made it to Swiftcurrent, the highest lookout in the park. It is a long, hard hike from The Loop, Logan Pass, or Swiftcurrent/Many Glacier to Swiftcurrent Lookout. Melanie Stine is pictured here. (Clayton J. Whitesides.)

Porcupine Ridge Lookout from the Kootenai Lakes. Porcupine Ridge Lookout is visible as a small dot on the ridge near the center-top of this photograph taken from the Kootenai Lakes, located south of Waterton Lake. In this faraway photograph, the isolation of and difficulty in accessing the lookout can be truly appreciated. Most of the materials used in construction of the lookout were brought to the site on horseback by legendary Glacier Park wrangler Bill Yenne.

Porcupine Ridge Lookout, 1940. This image shows Porcupine Ridge Lookout in its first season of use in 1940. Porcupine Ridge Lookout is a variation on the style of lookouts built at Apgar, Mount Brown, and Loneman. The catwalk was not as effectively supported, and the roof is at a gentler angle but does not extend out over the catwalk as at the Scalplock, Huckleberry, and Numa Ridge Lookouts. (GNP.)

Porcupine Ridge Lookout, 1973. At this point, the lookout was unmanned and clearly in poor shape; note the catwalk railing bowed out on the left. The lookout continued to deteriorate over the next 30 years. (GNP; photograph by Dave Shea.)

Deteriorated Porcupine Ridge Lookout, 2003. By 2003, much of the lookout's catwalk railing had collapsed, and the stairway was barely intact. The roof and siding were also in poor shape. This same year, the lookout was renovated because of its strategic position close to the Canadian border and near the Goat Haunt Ranger Station on the south shore of Waterton Lake, allowing the lookout to serve as a communications relay between Goat Haunt and park headquarters in West Glacier. (NPS.)

End Stages of Rehabilitation of Porcupine Ridge Lookout, 2003. The park service renovation crew relaxed as the rehabilitation of the lookout neared completion in the summer of 2003. The spectacular scenery from the lookout includes a view of the park's highest peak, Mount Cleveland, to the east. (NPS.)

HEAVEN'S PEAK AND THE BURN OF 1936. The results of the Heaven's Peak fire of 1936 are apparent in this view from Going-to-the-Sun Road near The Loop across McDonald Creek to Heaven's Peak. The 1936 fire was a primary spur in the construction of Heaven's Peak Lookout on the far end of the ridge leading to the right from the main peak. (National Archives; photograph by Ansel Adams.)

CONSTRUCTION OF MASONRY BASE OF HEAVEN'S PEAK LOOKOUT, 1945. In this photograph, Mennonite conscientious objectors are constructing the stone base of the Heaven's Peak Lookout during the summer of 1945. A contractor was hired to build the lookout beginning in August 1942, but the contractor claimed that wartime labor and material shortages prevented him from doing the job. He pleaded to be released from his contract and was released from it in December 1944. (GNP.)

Construction of Heaven's Peak Lookout Frame Cabin, 1945. The stone base serves as the cellar for the lookout as workers construct the frame of the main floor. Construction on the lookout was closed down by snowstorms in September 1945 when it was 95 percent complete, with no evidence that the final five percent of work would ever be done. (GNP.)

Helicopter View of Heaven's Peak Lookout, 1996. This bird's-eye view shows the positioning of the lookout on the northern tip of the ridge that extends from Heaven's Peak. The flat roof of the lookout had a hole in it when this photograph was taken.

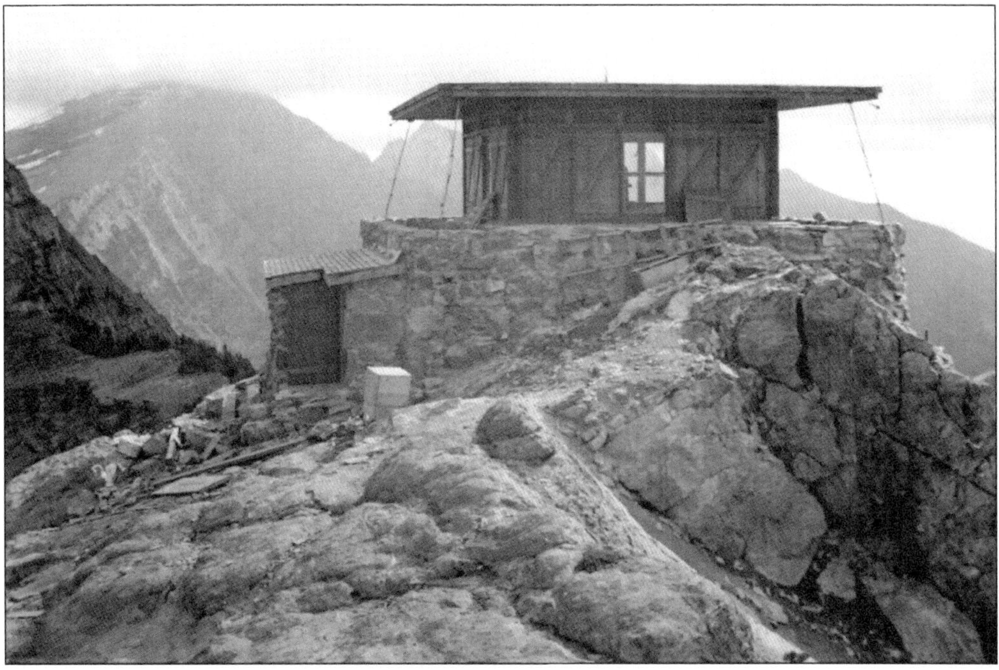

Heaven's Peak Lookout, 1951. The lookout, which was only manned from 1946 to 1953, is shown here during the latter part of its use. It had been reshingled in 1949 after snow and wind damage to the roof. Because of its isolated location and the difficulty in accessing it, the lookout was one of the first victims of the 1954 change in park policy that reduced the number of manned lookouts in favor of increased aerial observation. (GNP.)

Heaven's Peak Lookout Inventory Photograph, 1982. In this aerial inventory photograph, the lookout appears to still be in reasonably good shape. The next two decades, however, were rough on the structure. (NPS.)

Heaven's Peak Lookout, 1989. Rob Chaney, a reporter from the local newspaper, the *Hungry Horse News*, visited the lookout in August 1989. He described its "heavenly perch" and outstanding scenic views of the Camas Creek drainage and Longfellow Peak to the west and the Garden Wall and Going-to-the-Sun Road across the McDonald Creek drainage to the east. (HHN; photograph by Rob Chaney.)

Heaven's Peak Lookout on North Ridge of Peak, 2009. With a telephoto lens, Heaven's Peak Lookout is clearly visible from The Loop on Going-to-the-Sun Road. By 2009, the shutters on the windows were no longer in place, and the lookout was in a serious state of disrepair.

Heaven's Peak Lookout from Below, 2009. Members of the Glacier Mountaineering Society visited the abandoned lookout in 2009 and took numerous photographs of its deteriorated condition. This image offers a view of the lookout from below as one approaches the ridge crest. (Tim Anderson.)

Abandoned Heaven's Peak Lookout, 2009. The stone catwalk was littered with debris from inside the lookout when the Glacier Mountaineering Society visited it in 2009. Why someone would have extracted the bedsprings from inside the lookout and placed them on the catwalk is a mystery. (Tim Anderson.)

Interior of Abandoned Heaven's Peak Lookout, 2009. The debris in the interior of the lookout emphasizes its decaying state in 2009, as do boards pulling out from walls and the absence of the window shutters. (Tim Anderson.)

View through Windowless Heaven's Peak Lookout, 2009. Even with the lookout in its deteriorated condition, the scenery beyond the windows suggests what a special place Heaven's Peak Lookout is. Renovation work was scheduled to begin in 2011, but deep snow at the site, plus needed repairs elsewhere in the park—including at Scalplock Mountain Lookout—delayed the work until 2012. (Tim Anderson.)

New Shutters and Rehabilitation Supplies, Heaven's Peak Lookout, 2012. New shutters adorn the lookout after repairs by the NPS in the summer of 2012. Leftover supplies, a ladder, and the old bedsprings still needed to be cleaned up when the Glacier Mountaineering Society group visited the lookout that summer. (Tim Anderson.)

Heaven's Peak Lookout After Rehabilitation, 2012. Members of the Glacier Mountaineering Society, from left to right, Rick Murphy, Patti Markuson, Kim Davis, Dennis Twohig, April Suebert, and Tim Anderson, pose in front of the newly rehabilitated lookout in summer 2012. Old debris in the lower left corner of the photograph is waiting to be packed out. (Tim Anderson.)

Four

Former Glacier Park Lookouts

Several Glacier Park lookouts no longer exist; some were simply cabins on high points, such as the cabin on Riverview Mountain. A frame cabin on the flank of Heaven's Peak, known as "Heaven's Peak No. 1" or "Heaven's Peak South Lookout," was located in the middle of the slope on the Glacier Wall, two miles east of Heaven's Peak, and built in either 1943 or 1945 and destroyed in 1963.

Several "emergency lookouts" from the mid-1930s are no longer in existence or use; these include an elevated pole platform near the Many Glacier Hotel (the Alton Ridge Lookout), a high point with a map table and cabin overlooking Waterton Lake (the Goat Haunt Lookout), and the "Glacier Park Garage" emergency lookout, which was an observation point on the roof of a garage in East Glacier Park.

An incongruous steel tower lookout was placed in Red Eagle Meadow in 1959 and subsequently sold and removed in 1986.

Four two-story lookouts similar to those still in existence today have been torn down. The oldest, the Reynolds Ridge Lookout, was built in 1931 and burned down by rangers Bob Frauson and Charles Budge on November 28, 1962. The Elk Mountain Lookout, located near the southern edge of Glacier Park, immediately north of US Highway 2, was burned down by rangers Robert Sellers and Lou Hendrickx on May 27, 1965. Curly Bear Lookout (sometimes spelled Curley Bear in Glacier Park reports) was built in the late summer of 1934 with help from a CCC crew and was last used in 1959; in April 1963, rangers Frauson and Budge skied to the lookout and burned it down. The Bear Mountain Lookout (also known as the Crossley/Cosley Ridge) was built in 1935 to provide coverage in the area opened up by the new Chief Mountain International Highway. Bear Mountain Lookout was last manned in 1953; in April 1966, rangers Robert P. Wood and Fred J. Wood made a three-day ski patrol into Belly River and razed the lookout.

These lookouts were built by the NPS in the 1930s and fell into disuse in the 1950s, victims of changing fire-watching policy that favored aerial observation.

Bear Mountain Lookout. Bear Mountain Lookout, constructed in 1935, was a standard two-story cabin lookout with a steep, short roof similar to those at Apgar, Mount Brown, and Loneman Lookouts. It was located at an elevation of about 6,300 feet on the spur of Bear Mountain above Cosley (or Crossley) Lake and the Mokowanis River valley. (GNP.)

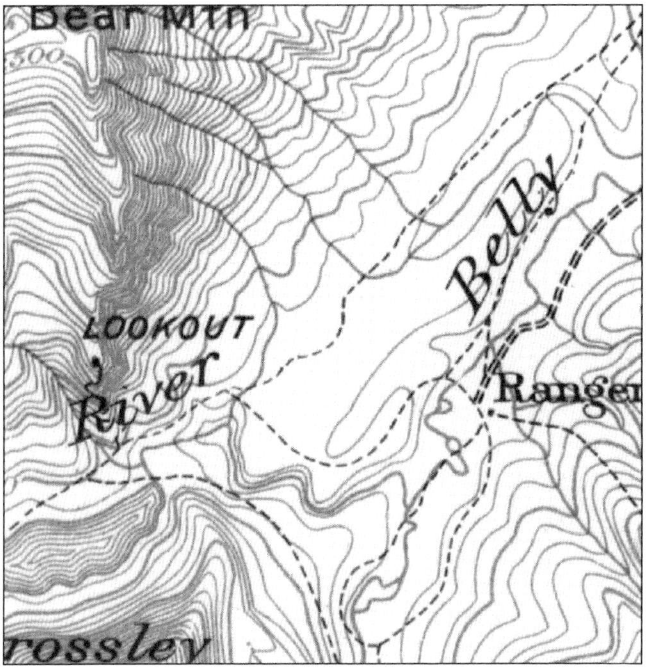

Topographic Map of Trail to Bear Mountain Lookout, 1938. The location of the former Bear Mountain Lookout, on a point extending out from Bear Mountain, is shown on this topographic map of Glacier Park. The Belly River Ranger Station, located at the end of the double-dashed former jeep road along the Belly River, was connected to the lookout by a primitive thick-strand-and-ceramic-insulator phone line similar to the one at Numa Ridge Lookout. (USGS.)

Historic Belly River Ranger Station, 2009. The Belly River Ranger Station is a historic structure from which supplies were provided to the Bear Mountain Lookout via trail. (Jon J. Kedrowski.)

Trail to Bear Mountain Lookout. The trail from the river valley up the steep slopes to the Bear Mountain Lookout passes through an ancient grove of pine trees and provides a tremendous view of Gable Mountain to the east. (Jon J. Kedrowski.)

Bear Mountain Lookout, 1950. The Bear Mountain Lookout, shown here in 1950, developed a weathered, sandblasted look because of the strong winds that blow down the Mokowanis River valley from the Continental Divide to the southwest. (GNP; photograph by R.A. Nelson.)

Panoramic View from Bear Mountain Lookout, August 1935. The Belly River valley, Chief Mountain (the peak farthest to the left), and the bulk of Gable Mountain dominate this photograph taken by Lester M. Moe in 1935. Note the corner of the catwalk railing in the lower right corner of the photograph. (GNP; photograph by L.M. Moe.)

Removing Debris from Razed Bear Mountain Lookout Site, 1973. After the lookout was destroyed in the spring of 1966, debris from the lookout was packed out in August 1966. Some of the heavier remaining wreckage was subsequently flown out in July 1973 by famous Glacier Park helicopter pilot Jim Kruger. The square shape of the lookout's foundation is visible below and to the right of the helicopter. (HHN; photograph by Mel Ruder.)

Base and Bolt at Site of Former Bear Mountain Lookout, 2009. A bit of the concrete base still contains a bolt at the site of the former lookout. Views from the former lookout site are still outstanding. (Jon J. Kedrowski.)

Reynolds Ridge Lookout Under Construction, 1931. The Reynolds Ridge Lookout (also occasionally referred to as the Mount Reynolds Lookout) was constructed in 1931 and first manned in 1932. Its final season of use was in 1953, and it was burned down by rangers Bob Frauson and Charles Budge in November 1962. (GNP.)

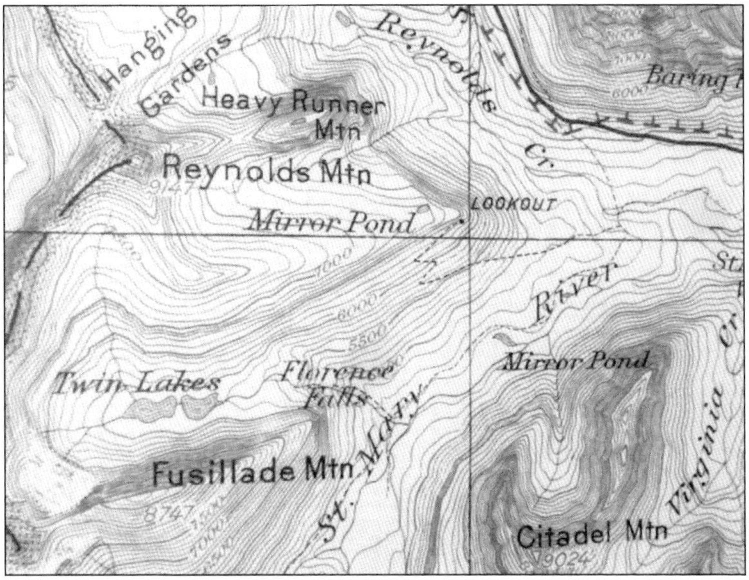

Topographic Map of Former Trail to Reynolds Ridge Lookout, 1941. The trail to the lookout, easily accessible from Going-to-the-Sun Road (upper right corner), climbed steeply from the floor of the St. Mary River valley to a ridge near Mirror Pond. This trail has fallen into disuse, and only minor fragments of it can still be found. (USGS.)

Husband-and-Wife Fire Watcher Team. Married firewatcher teams worked in both the FNF, such as the unidentified couple shown here, and at several Glacier Park fire lookouts, including at Reynolds Ridge. Both were trained to use the Osborne Firefinder and to report fire locations over radio and telephone. (USFS; photograph by K.D. Swan.)

Female Fire Watchers in Glacier Park. One of the advantages of husband-and-wife firewatcher teams was that one could hike out for mail or supplies (or a day off) while the other remained behind to watch for fires. In 1951 at Reynolds Ridge, Robert Bloedel and his wife saw lightning strike a snag about a quarter-mile from the lookout; Robert left to put out the fire, while Mrs. Bloedel remained behind to report its location. The unidentified firewatcher shown here worked at an unspecified fire lookout in the FNF. (USFS.)

Reynolds Ridge Lookout in 1949. Visits from backcountry rangers, including the one who took this and the following picture, were welcomed by fire watcher couples, as rangers would deliver much-appreciated mail and other reading materials to the isolated watchers. Most supplies, however, were typically delivered by packer and mule train, a tradition that continues today. (GNP; photograph by Adolph Opalka.)

Fire Watcher Couple at Reynolds Ridge Lookout, 1949. Some couples served only one summer at a fire lookout, while others returned several times. Not all lookouts in Glacier were "couples" lookouts, but Reynolds Ridge, Curly Bear, and Mount Brown in particular usually had couples occupying them. More isolated lookouts, such as Bear Mountain and Elk Mountain, were not typically deemed couples lookouts. (GNP; photograph by Adolph Opalka.)

Panoramic View to the Southwest/West from Reynolds Ridge Lookout, 1935. Mount Jackson and Jackson Glacier are visible at left from the former lookout. The trail leading up from the valley is shown here, and the outhouse facility is at far right, beyond the horses. (GNP; photograph by L.M. Moe.)

Panoramic View of Going-to-the-Sun Road from Reynolds Ridge Lookout, 1935. Going-to-the-Sun Road is visible in this panoramic view as the line running from upper left to center right. The fear of tourist-caused fires along the road was one of the driving forces for building this particular lookout. (GNP; photograph by L.M. Moe.)

Trail to Former Elk Mountain Lookout Site, 2009. This trail to Elk Mountain led to the site of the former lookout atop the peak. The site was praised for a lookout in a 1930 memorandum from fire chief George Paige to Glacier Park superintendent J. Ross Eakin, who echoed Paige's opinion in letters to the supervisor of the adjacent Flathead National Forest and to National Park Service fire control expert J.D. Coffman. A trail was built to the summit (elevation 7,835 feet) in 1930. (Jon J. Kedrowski.)

Undated View of Elk Mountain Lookout. Elk Mountain Lookout's design was similar to those of the Scalplock Mountain, Huckleberry Mountain, and Numa Ridge Lookouts (as well as the destroyed Reynolds Ridge and Curly Bear Lookouts). It was last used in 1959 and burned down by rangers Robert Sellers and Lou Hendrickx in 1965. (GNP.)

Remains of Base of Elk Mountain Lookout, 1996. From left to right, hikers Vern Ingraham (mostly hidden), Walt Bahr, Ed Prach, and Ivan O'Neil stand atop the remnants of the concrete base of Elk Mountain Lookout in 1996. The trail to the former lookout site is one of the first trails to become snow-free in the spring because of its southwest exposure, offering a good early-season hike. (HHN; photograph by Brian Kennedy.)

Debris from Elk Mountain Lookout and View to Northeast, 2009. The remote Ole Creek valley lies below the remnants of the former lookout. (Jon J. Kedrowski.)

US HIGHWAY 2 AND BNSF RAILROAD TRACKS FROM ELK MOUNTAIN, 2009. Looking south and southwest from the former lookout site offers views of US Highway 2 and the tracks of the Burlington Northern & Santa Fe Railway (originally the Great Northern Railway in this area). This image illustrates how high the lookout was above the surrounding terrain. Glacier Park is to the right of the railroad tracks, and the Flathead National Forest is in the distance at left. (Jon J. Kedrowski.)

CURLY BEAR LOOKOUT AND VIEW OF ST. MARY LAKE. Curly Bear Lookout, located high above the St. Mary valley, offered expansive views of St. Mary Lake and Singleshot Mountain. Much of the forest below and to the left of the lookout burned in the Red Eagle Fire of 2006. Curly Bear, a standard two-story cabin lookout with an extended roof over the catwalk, was last used in 1959 and burned down in 1963. (GNP.)

CURLY BEAR LOOKOUT ON 1960S "PIC-TOUR MAP" OF GLACIER NATIONAL PARK. Curly Bear Lookout, with its alternate "Curley Bear" spelling, was deemed sufficiently important to be included on this tourist-oriented Pic-tour Map from the late 1950s or early 1960s. The actual lookout looked nothing like its depiction on the map. (Don Bloodgood.)

PANORAMIC SOUTHWEST VIEW FROM CURLY BEAR LOOKOUT, 1935. St. Mary Lake is at the right edge of this photograph, with Red Eagle Lake near the center. The Red Eagle Fire of 2006 began near the head of that lake and burned much of the forest shown in the foreground. (GNP; photograph by L.M. Moe.)

PANORAMIC VIEW TO EAST/SOUTHEAST FROM CURLY BEAR LOOKOUT, 1935. Curly Bear Mountain blocks much of the view to the east/southeast from the lookout, but Divide Mountain is visible as the pyramid-shaped peak in the upper left background. Sharp eyes may be able to detect the dot of the Indian Service Divide Mountain Lookout ruins on the left-hand ridge of Divide Mountain. (GNP; photograph by L.M. Moe.)

MANY GLACIER HOTEL AND OLD ALTON RIDGE LOOKOUT SITE, 1921. The Alton Ridge Lookout was simply an elevated platform reached by a ladder and surrounded by a railing. The lookout was referred to in National Park Service correspondence in 1934 and shown on a 1937 "Seen Areas" fire map, but is otherwise not referred to anywhere in Glacier National Park archives. It was located to the right of the Many Glacier Hotel beneath Allen Mountain (at the right edge of the picture). (NPS; photograph by F. Kaiser.)

PANORAMIC VIEW TO SOUTHWEST AND WEST FROM ALTON RIDGE LOOKOUT, 1935. The Alton Ridge Lookout provided a spectacular view of, from left to right, Mount Gould, Grinnell Point, and Mount Wilbur, as well as Josephine (left) and Swiftcurrent (right) Lakes. Grinnell Glacier and Salamander Glacier are visible at the head of the valley at upper center, beneath the Garden Wall. (GNP; photograph by L.M. Moe.)

MANY GLACIER HOTEL COMPLEX VIEWED FROM ALTON RIDGE LOOKOUT, 1935. The Many Glacier Hotel sits on the shore of Swiftcurrent Lake in this panoramic photograph. The bulk of Altyn Peak looms behind the hotel, with Appekunny Mountain stretching off to the east (right). (GNP; photograph by L.M. Moe.)

View of Many Glacier Hotel Complex from Ground Level, 2007. Even without the benefit of the lookout platform (it is not known when it was removed), the view to the Many Glacier Hotel remains similar to the panoramic view from 1935. The top of Mount Henkel rises behind Alytn Peak in this photograph taken by the author.

Fire Map Location Table at Goat Haunt Lookout Site, 1935. The Goat Haunt Lookout consisted of a cabin and perhaps a low platform in addition to the fire map table shown here. The period of use and date of abandonment for this emergency lookout site are not recorded; however, it shows up on a park "Seen Area" fire map from 1937. The trail to this site, which starts from the Goat Haunt Ranger Station, remains intact. (GNP; photograph by L.M. Moe.)

Waterton Lake from Goat Haunt Lookout Site, 2006. Most of Waterton Lake and the surrounding forest are visible from the Goat Haunt overlook, site of the former emergency lookout. (NPS; photograph by David Restivo.)

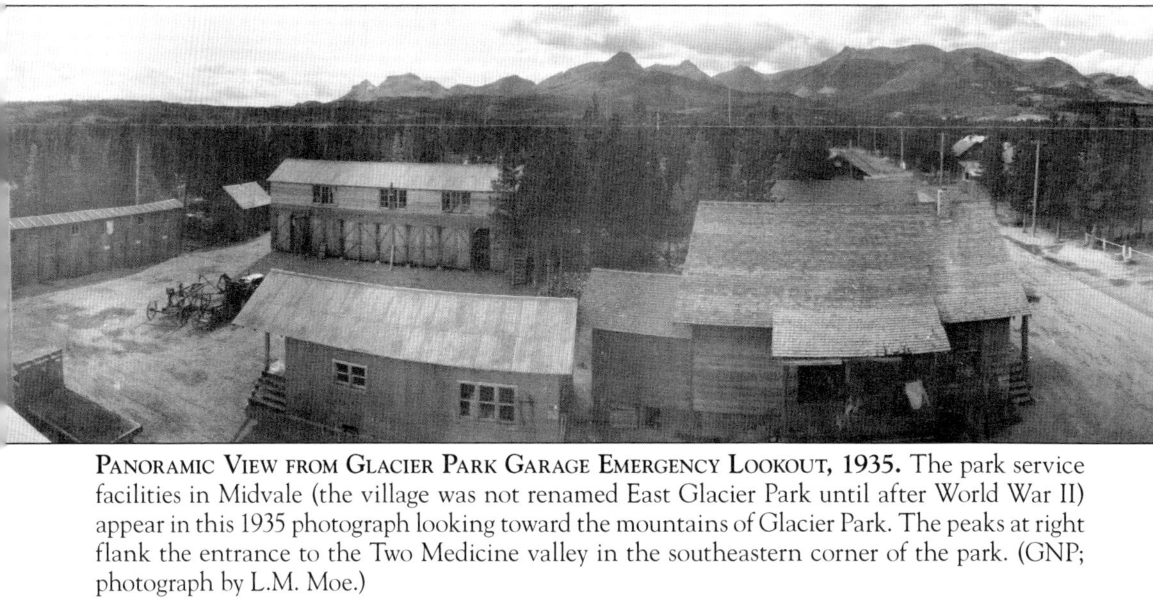

Panoramic View from Glacier Park Garage Emergency Lookout, 1935. The park service facilities in Midvale (the village was not renamed East Glacier Park until after World War II) appear in this 1935 photograph looking toward the mountains of Glacier Park. The peaks at right flank the entrance to the Two Medicine valley in the southeastern corner of the park. (GNP; photograph by L.M. Moe.)

Glacier Park Lodge Complex from Garage Emergency Lookout, 1935. The Glacier Park Lodge complex is visible near the upper right in this panoramic photograph taken from the roof of the Garage Emergency Lookout. (GNP; photograph by L.M. Moe.)

East Glacier Fire Truck Garage. Because of the changes in structures in East Glacier, it is difficult to know for certain which garage served as the base of the emergency lookout, but the pitch of the roof of this historic fire truck garage is similar to the one shown in the previous photograph. (NPS.)

Glacier Park Transport Company Garage, East Glacier, 1974. Another large garage near the complex in East Glacier is this garage that housed red "jammer" buses operated by Glacier Park Inc. Although the pitch of the roof is not the same as the one shown on page 89, this historic garage has long been associated with Midvale/East Glacier Park.

Base of Red Eagle Meadow Lookout, c. 1983. The most recent—and, to date, the last—lookout structure in Glacier Park is this 60-foot-tall steel tower topped with a seven-by-seven-foot room. The structure was on the Blackfeet Indian Reservation (see chapter five), before being moved by the NPS to the meadow overlooking St. Mary Lake in May 1959. It was first used at this site in July 1959. (GNP.)

Red Eagle Meadow Lookout, c. 1983. The Red Eagle Meadow Lookout was apparently only manned through the 1960s, during which time it was also used for winter observation of elk herds—two sleeping bags were placed in the tower in November 1962 for that purpose. The tower was clearly visible from Going-to-the-Sun Road. (GNP.)

Red Eagle Meadow Lookout and Divide Mountain, c. 1983. Red Eagle Meadow Lookout was considered visually incongruous with the natural setting of the area, according to park policies of the 1970s and 1980s. The lookout tower was auctioned off to a single bidder for $111 in June 1986 and removed in October of that same year. The crest of Divide Mountain is visible at lower left in this photograph. (GNP.)

Divide Mountain from Red Eagle Meadow, 1998. The only remaining evidence of the Red Eagle Meadow Lookout are two grown-over tracks that allowed access to the top of the ridge during the period when the Red Eagle Trail was still a jeep road for its first three miles or so. The meadow crest where the lookout was located still offers an excellent view of Divide Mountain, as shown in this photograph taken by the author.

Five

Blackfeet Indian Service Lookouts

The Blackfeet Indian Reservation lookouts were constructed in cooperation with Glacier National Park and representatives of the National Park Service. The oldest such lookout was constructed on a ridge of Divide Mountain in the early 1930s after park service fire expert J.D. Coffman, fire chief George Paige, and Francis X. "Frank" Guardipee of the Blackfeet Indian Reservation visited the mountain in August 1930. Following that visit, Blackfeet Indian Reservation superintendent Forrest Stone stated that he believed he would be able to secure funding for the Divide Mountain Lookout in fiscal year 1932. By 1934, the lookout had already been abandoned after being damaged by storms and high winds, along with dangerous lightning exposure. A nearby steel tower lookout, the Hudson Bay Divide Lookout, blew over in the winter of 1937, and Lester M. Moe took his panoramic photographs from atop the ruins of that tower in the summer of 1937.

The Looking Glass Ridge Lookout was built in 1937 on a long, flat ridge on the Blackfeet Indian Reservation at a point (elevation 6,288 feet) with a spectacular view that overlooks Two Medicine Valley. The lookout had a concrete base overlaid by a frame cabin with a catwalk. The winds on Looking Glass Ridge are notoriously strong, often exceeding 100 miles per hour during winter chinooks. Part of the cabin was blown away in 1940, and the lookout was subsequently abandoned. The graffiti-laden concrete base remains, as do some boards, old nails, and cables.

Two former lookouts existed on Swiftcurrent Ridge, approximately two miles west of Babb, north of the road into Many Glacier. In the 1930s, a frame cabin known as the Many Glacier Lookout existed at this site; the cabin was abandoned at an unknown date. During World War II, a 60-foot-tall steel tower, known as the Swiftcurrent Ridge Lookout, was erected on approximately the same site where the abandoned cabin was located. Although its exact date of installation is unknown, E.M. Hutchinson of Glacier Park noted the following on the "Seen Area" map of park lookouts: "I.S. (Indian Service) Swiftcurrent L.O. added 3/25/44." The Swiftcurrent Ridge Lookout was removed in 1959 and moved to Red Eagle Meadow inside Glacier Park.

Location of Former Swiftcurrent Ridge Lookout. The former Many Glacier Lookout and the former Swiftcurrent Ridge Lookout were both located at the top of the ridge in the center of this photograph. The ridge top provides a commanding view into the Swiftcurrent Valley to the west.

Southwestern View up Swiftcurrent Valley from Swiftcurrent Ridge, 1911. In 1911, when the US Geological Survey took this photograph, the dam impounding Lake Sherburne had not yet been constructed. At the head of the valley, The Salamander was still part of Grinnell Glacier, as they did not become separate ice masses until 1926. (USGS.)

LAKE SHERBURNE AND SWIFTCURRENT VALLEY FROM SWIFTCURRENT RIDGE LOOKOUT, 1954. In 1954, the US Geological Survey revisited the same general site on Swiftcurrent Ridge but used the Swiftcurrent Ridge Lookout to provide a higher vantage point. The Lake Sherburne dam and reservoir are visible at center. (USGS.)

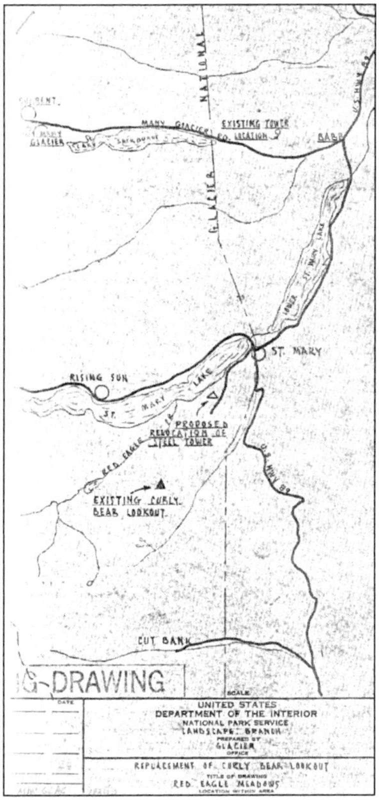

National Park Service Map Showing Proposed Relocation of Swiftcurrent Ridge Lookout to Red Eagle Meadow. In 1959, the park service proposed moving the Swiftcurrent Ridge Lookout to Red Eagle Meadow to replace the Curly Bear Lookout. The circle near the top of the map that shows the "existing tower location" is almost directly west of the town of Babb, and an empty triangle near the center of the map marks the "proposed relocation of steel tower" south of St. Mary Lake. The park service moved the tower in May 1959. (GNP.)

Divide Mountain Lookout, 2006. Divide Mountain Lookout is visible high on a ridge extending northward from the bulk of Divide Mountain, which is out of frame to the left of the photograph. There is no longer a real trail to the lookout.

RANGER FRANCIS X. (FRANK) GUARDIPEE. Guardipee, representing the Blackfeet Nation, helped guide park service personnel to the Divide Mountain ridge in 1930 and was integral in choosing the location for the Divide Mountain Lookout. Guardipee became a ranger at Glacier National Park in 1932 and served there until 1947. (NPS.)

DIVIDE MOUNTAIN LOOKOUT, 2011. Divide Mountain Lookout was an unusual octagonal structure atop a stone base. Cables hold the ruins in place against the strong winds on the ridge. The door has long been missing, and the ruins are now covered in graffiti. (Bob Sihler.)

Divide Mountain Lookout Ceiling, 1998. The unusual octagonal roof/ceiling of the Divide Mountain Lookout is shown in this photograph taken by the author. Note the metal cables designed to help hold the roof in place.

Repeat Photography Project at Divide Mountain Lookout, 2009. The US Geological Survey's repeat photography project, which replicates Lester M. Moe's 1935 and 1937 photographs from Glacier Park lookouts, came to Divide Mountain Lookout in 2009. Note the missing boards and generally precarious appearance of the lookout structure. (USGS, photograph by Chris Miller.)

Location of Former Hudson Bay Divide Lookout from Divide Mountain Lookout, 1998. The former Hudson Bay Divide Lookout was positioned on the knoll currently occupied by the structures and cellular phone tower shown here at the end of the unnamed gravel and dirt road. This photograph was taken from the Divide Mountain Lookout.

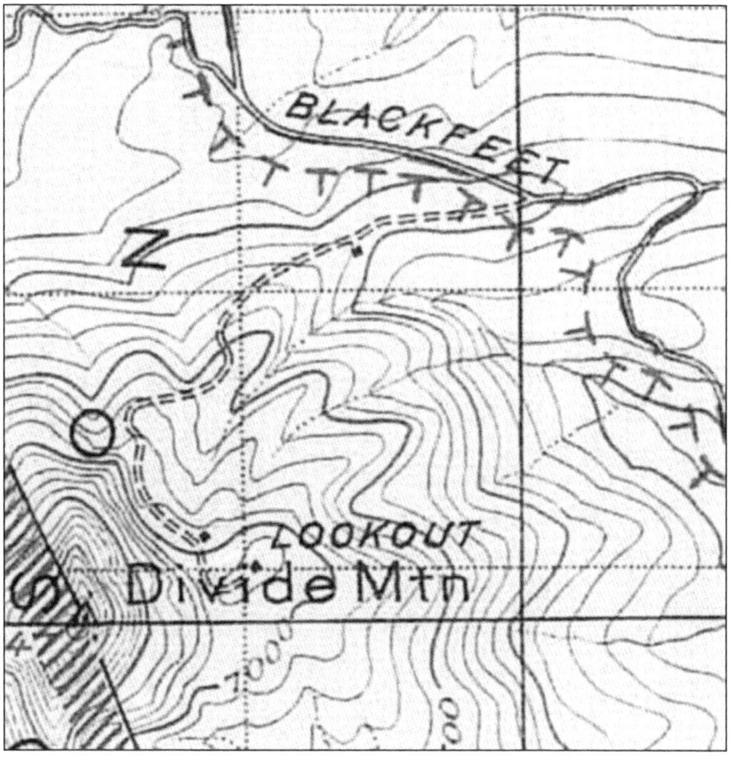

Topographic Map of Road to Hudson Bay Divide Lookout, 1941. The gravel-and-dirt road leading from US Highway 89 to the Hudson Bay Divide Lookout is shown on this 1941 topographic map of the eastern border of Glacier Park. The knoll upon which the steel tower was positioned is between the lowercase "i" and "d" in the word "Divide" on the map. (USGS.)

Modern Cellular Phone Towers at Site of Former Hudson Bay Divide Lookout, 2010. By 2010, an additional cellular phone tower had been erected on the knoll where the Hudson Bay Divide Lookout was once located.

Lester M. Moe and His Camera on Remains of Hudson Bay Divide Lookout, 1937. The steel tower lookout was constructed in 1934 and blown down in the winter of 1936–1937. Moe took his panoramic photographs from this precarious perch atop the ruins of the Hudson Bay Divide Lookout. (GNP.)

Panoramic View from Remains of Hudson Bay Divide Lookout, 1937. Lester M. Moe's photograph looks toward Divide Mountain (at center). Note Moe's automobile at left. A viewer using a magnifying glass to look at the original photograph could spot the Divide Mountain Lookout on the ridge leading down the right side of Divide Mountain. (GNP; photograph by L.M. Moe.)

View South from Looking Glass Ridge Lookout, 1921. The Looking Glass Ridge Lookout was built at this windswept site in 1937. Sixteen years earlier, the US Geological Survey used the excellent views from the same location to illustrate the peaks surrounding the Two Medicine Valley area. (USGS.)

Trees Showing Effects of Strong Winds on Looking Glass Ridge, 1921. The notoriously strong winds of Looking Glass Ridge are evidenced by the deformed shapes of these trees. Ice crystals carried by the strong winds trim buds from the windward sides of tree trunks, leaving only downwind, leeward branches; this gives the trees a lopsided, flag-like appearance. (USGS.)

Remains of Looking Glass Ridge Lookout, 1969. The remains of the Looking Glass Ridge Lookout are visible at the center of this photograph. The upper floor of the lookout was blown away in 1940. The remains were accessible via a gravel road leading eastward from Montana Highway 49, but that road was blocked by a mudflow in 2009 and is no longer passable. (GNP; photograph by John Tyers.)

Remains of Looking Glass Ridge Lookout, 2011. The concrete base of the Looking Glass Ridge Lookout is heavily laden with graffiti, as shown here. The view from the site is still spectacular. (Clayton J. Whitesides.)

Panoramic View of Two Medicine Valley from Looking Glass Ridge Lookout, 1937. Lester M. Moe's spectacular photograph is heightened by the use of black-and-white infrared film, which gives the view its unusual tones. Montana Highway 49, at lower left, had not yet been paved at the time the photograph was taken. (GNP; photograph by L.M. Moe.)

GRAFFITI ON REMAINS OF LOOKING GLASS RIDGE LOOKOUT, 1999. This view from the site of the Looking Glass Ridge Lookout, with Scenic Point the closest peak to the south, shows little change—other than the graffiti-covered remains—from the photograph taken by the US Geological Survey in 1921.

Six

Cooperating Flathead National Forest Lookouts

Early lookouts in the Flathead National Forest, like those in Glacier Park, were simply high points on peaks that offered expansive views of the surrounding forest. Also like in Glacier, map tables and phone lines were built for these sites, followed by the construction of cabin lookouts and, eventually, a variety of tower forms. There were 146 structures in the Flathead National Forest that made up a web of forest-fire detection, although most of these did not offer views of Glacier Park. Today, fewer than 15 lookouts remain in the Flathead National Forest, and only three remain adjacent to Glacier Park.

Fifteen US Forest Service lookouts existed in the 1930s along the western borders of Glacier National Park as part of a cooperative network of lookouts that provided reciprocal assistance, with National Park Service fire watchers reporting on fires in the Flathead National Forest and USFS watchers reporting on fires in Glacier Park that had not been spotted by their own respective lookouts. Most of the Flathead National Forest lookouts were positioned along the northwestern border of Glacier Park formed by the North Fork of the Flathead River. Because Glacier Park had several lookouts offering views of its southwestern border along the Middle Fork of the Flathead River, only four forest service lookouts served that area.

Today, most of the forest service lookouts have been removed. Cyclone Peak Lookout is still in use, and Thoma Lookout was recently renovated. Hornet Peak Lookout, the oldest still in existence in the area, is available for recreational rental from the US Forest Service. Remnants of Coal Ridge Lookout still exist, but the lookout itself has been long abandoned. As part of his Glacier Park project, Lester M. Moe took panoramic photographs from the lookouts that existed in the 1930s; for many of these lookouts, Moe's photographs are the only remaining evidence of the network that once protected both the eastern flanks of the Flathead National Forest and the western fringes of Glacier National Park.

Early Forest Service Rangers and Fire Watchers. Early 20th-century Flathead National Forest rangers, including Frank Liebig (at far right), pose in this photograph. Liebig's duty was to watch for fires and protect most of the western half of what became Glacier National Park in 1910. (USFS.)

Flathead Forest Fire Watcher, 1923. As in Glacier Park, early lookout sites in the Flathead National Forest consisted simply of high points offering expansive views of the surrounding terrain. This fire watcher has a map table with an early Osborne Firefinder upon it. The Osborne Firefinder—invented by William "W.B." Osborne, a US Forest Service employee—has been in use since 1915. (USFS.)

Fire Watcher Reporting Via Telephone, 1920s. This fire watcher is reporting on a single-line telephone in the Flathead National Forest. Such phone systems were slow to come to the lookout peaks and often required the fire watcher to hike down several miles from his map table/Osborne Firefinder in order to reach the phone and call in a fire's location. (USFS.)

Hornet Peak Lookout, 1923. Hornet Peak Lookout, built in 1922, replaced a pole tower and cabin built in 1915; the pole tower and older cabin are adjacent to the new lookout in this image. Hornet Peak Lookout is a roof cupola–style lookout. (USFS; photograph by Howard Flint.)

Modern View of Hornet Peak Lookout. The old pole tower and cabin are gone, but Hornet Peak Lookout still stands in the Whitefish Range. It is available for summer rental for those looking for an off-the-beaten-path visitor experience. (USFS.)

Thoma Lookout. Built in 1930, the historic Thoma Lookout was used during emergencies until 1973. It was reconditioned in 2010 and has since been manned each summer. A short documentary film directed by Brian Bolster, *The Lookout*, features Thoma Lookout; the film premiered at the 2012 Slamdance Film Festival in Park City, Utah. (Brian Bolster.)

BURNED LANDSCAPE NEAR THOMA LOOKOUT, 1935. This black-and-white infrared photograph from Thoma Lookout shows the effects of a recent forest fire. Dead trees around the lookout were chopped down to improve the view and to remove hazards such as weakened trees falling onto the trail or the lookout. (GNP; photograph by L.M. Moe.)

THOMA LOOKOUT ON A CLOUDY DAY. Thoma Lookout remains the lookout closest to the Canadian border adjacent to Glacier Park. It provides an excellent view of the peaks around the foot of Kintla Lake in the northwest corner of the park. (Brian Bolster.)

PANORAMIC VIEW INTO GLACIER PARK FROM CYCLONE PEAK LOOKOUT, 1935. The effects of a forest fire around Cyclone Peak Lookout—which is still in operation today—are obvious in this 1935 photograph. Bowman Lake, located in Glacier Park, is clearly visible from the lookout (and near the upper right corner of this picture), and careful scanning with binoculars may reveal Numa Ridge Lookout above the lake. (GNP; photograph by L.M. Moe.)

PANORAMIC VIEW INTO GLACIER PARK FROM DEMERS RIDGE LOOKOUT, 1935. This view from the Demers Ridge Lookout was similar to the one from nearby Huckleberry Lookout in Glacier Park. The Demers Ridge Lookout pole-tower structure was destroyed in 1957. (GNP; photograph by L.M. Moe.)

VIEW FROM DESERT MOUNTAIN LOOKOUT, 1935 AND 1956. Desert Mountain Lookout, located near the entrance to Glacier Park by West Glacier, overlooked a large part of the park headquarters and Apgar areas, as well as the surrounding Flathead National Forest. A pole tower was built on the mountain in 1930, and a road to the top of Desert Mountain was built by Civilian Conservation Corps workers in 1933. Panoramic photographs were taken from the lookout in both 1935 (above) and 1956 (below), allowing for an idea of how the landscape changed over the course of 21 years. Substantial regrowth from the Half Moon Fire of 1929, the effects of which were still very apparent in 1935, characterize the 1956 picture. The lookout was burgled in the early 1970s and destroyed in 1977. (Both, GNP; above, photograph by L.M. Moe.)

Panoramic View into Glacier Park from Ousel Peak Lookout, 1935. Ousel Peak Lookout was a cabin-style structure built near West Glacier in 1931. It offered expansive views of the Middle Fork of the Flathead River region and excellent views of Glacier Park's southern realm, from Mount Stimson to beyond Mount St. Nicholas, as shown here. The lookout, which was in the line of sight from Loneman Mountain Lookout in Glacier Park, was destroyed in 1957. (GNP; photograph by L.M. Moe.)

Garry Lookout, 1971. Garry Lookout (sometimes spelled Gary) was sometimes manned by personnel from Glacier Park in the 1960s, even though it was in the Flathead National Forest southeast of West Glacier. The 30-foot-tall pole-tower structure with a cabin on top was built in 1934. It offered excellent views into the Middle Fork region of Glacier Park and was destroyed in 1974. (GNP.)

Seven

Roadside Lookouts

In addition to the panoramic photographs taken from every fire lookout in Glacier Park, the Flathead National Forest lookouts overlooking Glacier Park, and the Blackfeet Indian Service lookouts along Glacier Park's eastern edge, Lester M. Moe took panoramic photographs from several roadside lookouts along Going-to-the-Sun Road west of Logan Pass. He also took several photographs from roads on the eastern edge of the park. Interestingly, he did not take photographs from Going-to-the-Sun Road east of Logan Pass, presumably because the forests along that stretch of road were monitored from the fire lookouts at Reynolds Ridge and Curly Bear.

Moe took the photographs along the western side of Going-to-the-Sun Road on July 19, 1935, with the exception of the one from the highest roadside lookout, which was taken on August 20, 1935, when he took all of the eastside photographs. The westside photographs provided coverage along Lake McDonald from roadside points one and two; much of the forest on the west side of Lake McDonald burned in the Robert Fire of 2003. The McDonald Creek valley was photographed from roadside point three. Roadside points four and five provided sweeping views along the road from McDonald Creek to The Loop and up to near the road's Triple Arches. Point six was photographed in August 1935 from between Triple Arches and Oberlin Curve, just below Logan Pass.

Views along the roads of the east side of the park provided coverage for areas largely unseen in views from fire lookouts, although the Two Medicine roadside photographs largely replicate those taken in 1937 from Looking Glass Ridge Lookout, which was not in existence in 1935. Photographs from US Highway 89 provided coverage of the Cut Bank valley area and the area near Divide Creek, south of the town of St. Mary. The final roadside lookout on the east side was along the road into the Many Glacier area from the roadside pullout adjacent to the Lake Sherburne Reservoir.

Roadside Lookout along McDonald Creek, 1935. The dense western redcedar forests of McDonald Creek valley were visible from Roadside Lookout No. 3, as labeled on Lester M. Moe's photographs. Mount Brown is visible as the mountain at right, and Going-to-the-Sun Road crosses most of the lower part of the image. (GNP; photograph by L.M. Moe.)

Roadside Lookout above The Loop Looking North, 1935. The Loop on Going-to-the-Sun Road and the surrounding forest could be seen from Roadside Lookout No. 5. Much of the forest around The Loop burned in the Heaven's Peak Fire of 1936 and again in the Trapper Fire of 2003. (GNP; photograph by L.M. Moe.)

ROADSIDE LOOKOUT APPROACHING LOGAN PASS FROM THE WEST, 1935. Roadside Lookout No. 6 offers a broader view of the area around The Loop. Haystack Butte and the Garden Wall loom above the road in the central background. Note the unpaved surface of Going-to-the-Sun Road. Keeping the road paved and in good condition in an area prone to rockfalls and snow avalanches is a constant challenge. (GNP; photograph by L.M. Moe.)

LESTER M. MOE'S CAR ON GOING-TO-THE-SUN ROAD, 1935. Moe's automobile is parked beneath a patch of snow at Roadside Lookout No. 6. The rocks and snow that can affect the road are readily apparent in this photograph. (GNP; photograph by L.M. Moe.)

BURNED TREES ALONG HIGHWAY 49 NEAR TWO MEDICINE AREA, AUGUST 1935. This image captured a beautiful day in the Two Medicine area as observed from the Two Medicine Roadside Lookout off of Montana Highway 49 (the Looking Glass Road). Rising Wolf Mountain, the highest peak in the Two Medicine valley, looms right of center. (GNP; photograph by L.M. Moe.)

LAKE SHERBURNE AND SWIFTCURRENT VALLEY, 1935. The Sherburne Lake Roadside Lookout, adjacent to the Lake Sherburne Reservoir, provided a sweeping view of the Swiftcurrent Valley and Many Glacier region. Grinnell Glacier and The Salamander are visible near the center background. (GNP; photograph by L.M. Moe.)

Eight

The 21st Century and Fire in Glacier National Park

From the founding of Glacier National Park in 1910 until the early 1980s, park policy was to attempt to put out every fire as soon as possible. It is now known that precluding fire from occurring produces an unhealthy build-up of fuel within the forest and on the forest floor, as well as removing the ability of fire to destroy pests such as the mountain pine beetle and spruce budworm, which have prospered in broad areas of Glacier Park's forests. Since the early 1980s, park personnel have recognized the importance of fire in the ecology of the forests of Glacier Park and surrounding areas. Some natural fires in remote locations are now allowed to burn, and prescribed fires have also been used on a limited basis to clear out heavy undergrowth. However, fires near tourist facilities are still suppressed as much as possible.

The build-up of fuel that occurred over more than 70 years of fire suppression left many areas in the park vulnerable to hot crown fires that can destroy thousands of acres of forest. Since the start of the 21st century, notable fires within the park included the Moose Fire in 2001; several fires in 2003 that collectively burned over 100,000 acres (more than 10 percent of the total Glacier Park area), including the Robert, Trapper, and Middle Fork Complex Fires; and the Red Eagle Fire of 2006. The Skyland Fire of 2007 burned in the forest just south of Marias Pass along the southern boundary of Glacier Park and threatened to jump US Highway 2 and enter the park. Smoke from that late July fire temporarily halted car traffic in the Marias Pass area.

Visitors to the park who observe the dead trees from a forest fire often complain that the landscape looks ravaged and unnatural. On the contrary, forest fires have always been and will continue to be natural parts of the Glacier Park landscape. The National Park Service now attempts to educate the public about the importance of natural fires in the Glacier Park ecosystem.

Robert Fire behind Apgar Mountain As Seen from West Glacier, 2003. The Robert Fire threatened West Glacier and caused its temporarily evacuation. Apgar Mountain (in the background) was heavily burned, but Apgar Lookout was covered in spray-foam retardant and survived the fire.

RESULTS OF 2003 ROBERT FIRE ON TRAIL TO APGAR LOOKOUT, 2007. The Robert Fire burned across most of the trail to Apgar Lookout. Dead trees continue to fall across the trail during times of high winds, such as shown here in front of James Dietrich. (Shannon Paprocki.)

Burned Sign along Camas Road in Glacier Park, 2003. The Robert Fire burned across the Camas Road north of Apgar, scorching a speed limit sign and burning through a forest of lodgepole pine trees. These lodgepole pines replaced a broad area burned in 1929.

New Sign and Regrowth along Camas Road, 2004. A new speed limit sign was installed along the same stretch of the Camas Road a year after the Robert Fire. The roadside vegetation had recovered, and new lodgepole pine seedlings were on the forest floor, marking the start of a new, post-fire forest.

Effects of 2003 Robert Fire in Mature Forest above Lake McDonald, 2004. The Robert Fire burned along the west side of Lake McDonald to near the head of the lake before being stopped. The fire burned into a mature forest along the trail to Trout Lake above Lake McDonald, scorching trunks, as shown here surrounding hiker Dawna Cerney. Many of these large trees survived, but others were weakened by the fire and subsequently died.

Warning Sign Due to Effects of the 2003 Robert Fire. The lasting hazards of a forest fire are illustrated on this warning sign photographed by the author in September 2004.

Effects of 2003 Trapper Fire As Viewed from The Loop, 2007. The Trapper Fire burned much of the forest below Heaven's Peak. The Heaven's Peak Lookout is located on the flat area on the right-hand end of the rocky ridge at upper left, barely visible here. (Shannon Paprocki.)

IMPACT OF 2003 TRAPPER FIRE ON THE TRAIL TO SWIFTCURRENT LOOKOUT, 2011. The trail to Swiftcurrent Lookout passed through a grove of low-growing trees at the upper tree line. Those trees, shown here, were burned in the Trapper Fire of 2003. Fire watcher Chris Baker, who was in Swiftcurrent Lookout during the fire, provided valuable information on the fire's situation. (Melanie B. Stine.)

SWIFTCURRENT LOOKOUT AND EFFECTS OF 2003 TRAPPER FIRE, 2011. The effects of the Trapper Fire are evident when looking toward the peak of Swiftcurrent Mountain, where Swiftcurrent Lookout is barely visible atop the center of the crest. Trees grow very slowly at this elevation, and it will be decades before another tree line forest reoccupies the site. (Melanie B. Stine.)

Effects of 2006 Red Eagle Fire in St. Mary Valley, 2012. The effects of the Red Eagle Fire are starkly apparent in this photograph. Singleshot Mountain dominates the upper part of the image. This fire, which began in Glacier Park near the head of Red Eagle Lake, would have been easily spotted from the Curly Bear or Red Eagle Meadow Lookouts had they still been in existence at the time of the fire.

EFFECTS OF 2006 RED EAGLE FIRE ON BLACKFEET INDIAN RESERVATION, 2007. Looking toward Divide Mountain from US Highway 89 reveals broad swaths of burned areas on the Blackfeet Indian Reservation. The Red Eagle Fire burned over 34,000 acres in Glacier National Park and on Blackfeet tribal land. The fire was first reported at approximately 1:00 p.m. on July 28, 2006. Timber harvesting was underway on the reservation in this image taken a year after the fire.

SKYLAND FIRE ON THE SOUTHERN BOUNDARY OF GLACIER PARK, 2007. As shown in this photograph taken from US Highway 2 in the Bear Creek canyon west of Marias Pass, the Skyland Fire was "blowing up" out of control on July 28, 2007.

Skyland Fire from Marias Pass, 2007. From this vantage point along Marias Pass on the Continental Divide, on the southern border of Glacier Park, flames could be seen rising 50 to 100 feet into the air on July 28. Soon after this photograph was taken by the author, traffic coming west from the pass into Bear Creek Canyon was temporarily halted because of the dense smoke.

Skyland Fire As Seen from Area Burned by the 2006 Red Eagle Fire. This image, taken from near the crest of Milk River Ridge on US Highway 89 south of St. Mary on July 28, 2007, shows the effects of the burning Skyland Fire and the previous summer's Red Eagle Fire. Portions of Highway 89 were melted by the heat of the Red Eagle Fire and had to be repaved.

BIBLIOGRAPHY

Frohlicher, S.V. *It Happened in Glacier Park: Bears, People 'n Other Stuff by an Oldtime Fireguard.* Great Falls, MT: Licini's Print, 1987.

Gildart, Robert C. *Montana's Early-Day Rangers.* Helena, MT: Montana Magazine Inc., 1985.

Glacier National Park Archives, West Glacier, MT.

Guthrie, C.W. *The First Ranger: Adventures of a Pioneer Forest Ranger, Glacier Country 1902–1910.* Huson, MT: Redwing Publishing, 1995.

Hufstetler, Mark. "Life on Porcupine Ridge: A Fire Lookout's Diary." *The Inside Trail (Voice of the Glacier Park Foundation)* 8(1) (1988): 6-7, 14-15.

Hungry Horse News, weekly newspaper from Columbia Falls, MT, numerous issues.

Kresek, Ray. *Fire Lookouts of the Northwest.* Spokane, WA: Historic Lookouts Project, 1998.

Ober, Michael J. "The CCC Experience in Glacier National Park." *Montana: The Magazine of Western History* 26(3) (1976): 30-39.

Parratt, Mark W. *Fate is a Mountain.* Whitefish, MT: Sun Point Press, 2009.

Peacock, Doug. *Grizzly Years.* New York, NY: Henry Holt and Company, 1990.

Robinson, Donald H. *Through the Years in Glacier National Park.* West Glacier, MT: Glacier Natural History Association, 1960.

Yenne, W.J. *Switchback.* Kalispell, MT: WY Books, 1983.

Discover Thousands of Local History Books Featuring Millions of Vintage Images

Arcadia Publishing, the leading local history publisher in the United States, is committed to making history accessible and meaningful through publishing books that celebrate and preserve the heritage of America's people and places.

Find more books like this at
www.arcadiapublishing.com

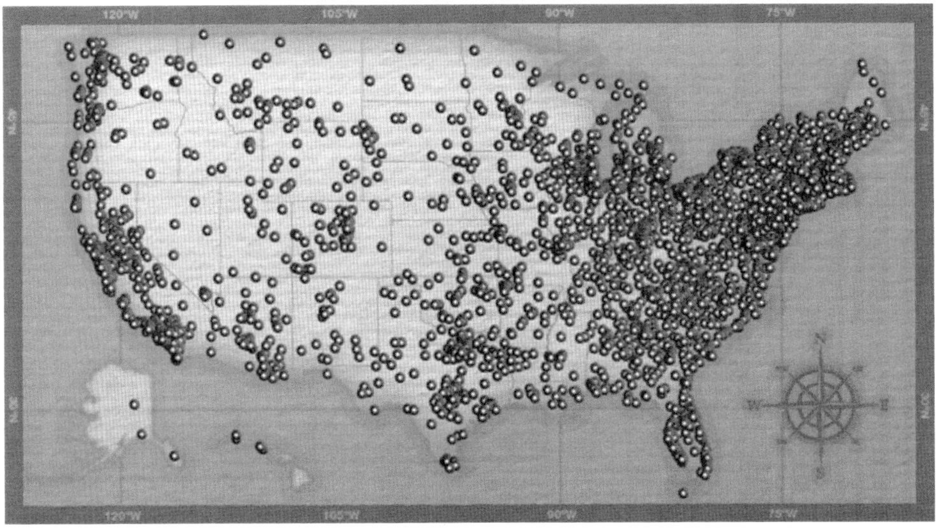

Search for your hometown history, your old stomping grounds, and even your favorite sports team.

Consistent with our mission to preserve history on a local level, this book was printed in South Carolina on American-made paper and manufactured entirely in the United States. Products carrying the accredited Forest Stewardship Council (FSC) label are printed on 100 percent FSC-certified paper.

MADE IN THE USA